FAIRBANKS-MORSE
LOCOMOTIVES In Color

BY JIM BOYD
Photography by the author, except as noted

Published by
Morning Sun Books, Inc.
11 Sussex Court
Edison, NJ 08820
Library of Congress Catalog Card Number: 96-076561
First Printing
ISBN 1-878887-63-7

Layout and Design by Jim Boyd
Photography by Jim Boyd, except as credited

Color separation and printing by

THE KUTZTOWN PUBLISHING CO., INC.

Kutztown, Pennsylvania

HAWK MOUNTAIN CHAPTER NRHS / MIKE DEL VECCHIO COLLECTION

Dedication

To Mike Del Vecchio, who some day will own a Lackawanna Train Master builders plate.

Acknowledgments

The author would like to thank the following people for their contributions and help in preparing this book: First, of course, is Bob Yanosey, for not only agreeing to publish the book but for digging into his vast collection for slides from himself and his regular contributors that helped so much in expanding the scope and completeness of the locomotive coverage. Thanks go to all whose material from Bob's collection was used, but one deserves special mention for the sheer quantity and quality of material: Matt Herson.

Supplying material directly to the author were Walt Appel; Keith Ardinger; Peter Bergs; Mike Del Vecchio; Frank Etzel; Bill Farmer; Kermit Geary, Jr., and Senior; Ken Goslett; John Lyle; Alan Miller; Jeremy Plant; Russ Porter; Kurt Reisweber; Mike Schafer; Dick Townley;

Dick Wallin and Don Wallworth, as well as Sandy Burton, Preston Cook, Tony Koester, Terry La France, Stan Smaill and Bob Wanner, who provided additional material. All photos from collections are also credited to the original photographers if the names are known.

Joe Van Hoerebeke and Jere Lee Geib supplied photos of FM builders plates from their collections, and Wayne Monger provided the list of surviving FM locomotives. *Extra 2200 South*, *The Short Line* and John Kirkland's book *The Diesel Builders, Fairbanks-Morse and Lima-Hamilton* (Interurban Press, 1985) were sources of roster information in addition to numerous original sources. And special thanks go to the folks at the Illinois Railway Museum for preserving Milwaukee Road 760 and making it available for photography. ⊕

PHOTOGRAPHY IN THIS BOOK BY JIM BOYD, EXCEPT AS OTHERWISE CREDITED

THE FIRST DIESEL I ever got to know was a Fairbanks-Morse switcher. I didn't know the make or model at the time because I was only eleven years old, but in 1952 everything else around home on the Illinois Central was steam — and that yellow and green thing on the North Western was a "diesel." I had seen "streamliners" fly through town on the C&NW main line, but in memory they were just a blur. The image of the dark green and shrouded 4-6-4 4002 taking on water with an eastbound "Clinton Local" was much more vivid. But that diesel switcher with a steam engine cab roof "lived" in Dixon and shared the street trackage along the Rock River with my ICRR 0-6-0s. The C&NW had dieselized the Dixon Job a few years before, and I don't remember ever seeing a North Western steam switcher. When I became "aware" of trains, it was through the IC Mikados and the resident 0-6-0 and the C&NW diesel switcher that ventured downtown — the C&NW double-track main line was far on the south side of town and for all practical purposes out of reach. The IC, however, passed within a block of my northside home, and the steam engines paraded grandly past in easy view.

But through the open summertime windows in my bedroom, I could hear the evening drama as the two switch engines — the IC 0-6-0 and the North Western diesel — would team up to doublehead the Medusa Portland Cement Company loads uphill out of the river valley to the main line yards on the south side. The steam engine would always lead, with its sharp exhaust and steamboat whistle clearing the crossings from River Street to Seventh Street — with the pleasantly drumming sound of the diesel filling in behind. The few times that I actually witnessed it, I remember the diesel accelerating beneath a plume of blue smoke.

By the time I got a camera in 1953, I wasn't about to waste film on diesels. My only snapshots are of IC steam and one stray C&NW 4-6-0 on a work train. By the time I began taking train pictures seriously, both the IC and C&NW had turned to Geeps to switch Dixon. All I remember is the number of the C&NW diesel that came to town every day from Nelson yard: 1082. I don't have any pictures of it.

I spent my teens trying to grasp the last of the departing steam locomotives, and it wasn't until the mid-1960s that I began to photograph diesels for their own sake. Not surprisingly, I always had a fond spot for the boxy diesels with their blue smoke that I came to know as the products of Fairbanks-Morse of Beloit, Wisconsin. But by the time I knew enough to truly appreciate them, they were nearly gone.

As I was gathering the photos for this book, I was surprised at how few pictures I had of the C&NW FM switchers that had been so much a part of my childhood. Though I didn't realize it at the time, this encounter (above) with the 1054 at Green Bay in August 1975 was the last I would have with the yellow and green locomotives that had brought the word "diesel" into my life.

Today, nearly all the FM's are gone. But join me here to relive some memorable FM encounters, both first-hand and through the photographic efforts of others — JIM BOYD, August 8, 1996.

FAIRBANKS-MORSE
AND THE LOCOMOTIVE BUSINESS

FAIRBANKS, MORSE & Company had been associated with railways from its earliest years. In 1830 Erastus and Thaddeus Fairbanks of St. Johnsbury, Vt., promoted the construction of railroads to serve their mechanical platform scale manufacturing business (Thaddeus was the inventor of the platform scale). In 1866, Charles Morse, a longtime employee, became a partner in the Cincinnati, Ohio, office, and the name soon changed to Fairbanks, Morse & Company, with Morse taking the leadership role from the Fairbanks family.

In 1885 Morse "diversified" from the scale manufacturing business by acquiring the Eclipse windmill company and the Beloit Wagon Works in Beloit, Wisconsin. Windmills drove water pumps, and replacing windmills with steam and gasoline engines was a logical step in a company's development — and steam locomotives needed lots of water, which could be supplied by FM pumps and engines. By acquiring the Sheffield Velocipede Car Company, Morse got into the railway supply business, making handcars and then gasoline-driven speeders for track maintenance workers. Pump-driving engines could also be employed in lifting coal into locomotive fueling stations, and Fairbanks-Morse became a manufacturer of huge railroad coaling towers.

Seeking better power plants for its stationary facilities, FM got into the diesel engine business in 1922 by hiring Fritz P. Grueztner, a senior engineer with the De La Vergne Engine Company (which ultimately powered Baldwin's first diesel locomotives). Thus Fairbanks-Morse entered the 1930s with a line of two- and four-cycle diesel engines of varying size and horsepower.

The success of the German "U-boats" during the First World War had the U.S. Navy seeking a lightweight power plant for submarines. Fairbanks-Morse survived the Great Depression by developing a distinctive opposed-piston, two-cycle diesel engine which met the Navy's specifications at the time when General Motors was still experimenting with its historic Winton 201A two-cycle engine.

In 1939 FM supplied five-cylinder opposed-piston (OP) engines rated at 800 h.p. for six streamlined diesel-electric motor cars being built by the St. Louis Car Company for the Southern Railway. These shovel-nosed baggage-mail/locomotive units (passengers rode in an unpowered trailer coach) were the first application of the OP engine to a piece of railroad equipment. The cars had a distinctive 2-A1A wheel configuration with two unpowered axles under the cab and a three-axle truck with two outboard traction motors and center idler under the rear. Southern 1-4 were used on the Birmingham-Mobile *Golden Rod*; Oakdale, Tenn.,-Tuscumbia, Ala., *Joe Wheeler* and Atlanta-Brunswick, Georgia, *Cracker*, while the Alabama Great Southern 40 and 41 covered the *Vulcan* schedule between Chattanooga, Tenn., and Meridian, Mississippi. Of these, No.2, of the *Joe Wheeler*, wound up on the Georgia Northern and survived into the 1970s; sometime during its career, a non-powered three-axle truck replaced the two-axle truck under the cab. The car was offered for preserva-

tion in 1966, but bad wheels and the cost of transportation to an Atlanta museum doomed it to oblivion.

A side-note to FM's history, however, is its first diesel application in a true locomotive: Reading Company No.35 (later renumbered 97). This ill-fated centercab switcher built by the St. Louis Car Company in 1939 was powered by two eight-cylinder, four-cycle 300-h.p. inline engines, one under each hood. The locomotive's downfall was that the two diesels were each connected to one auxiliary generator beneath the cab floor. The two unsynchronized engines "fought" each other through that auxiliary generator, vibrated excessively and produced the highest maintenance-cost internal combustion locomotive ever to grace the Reading roster — it remained for its entire career at the Erie Avenue Yard in Philadelphia and was finally traded on a Train Master in 1953.

Following the 1939 Southern motor cars, the U.S. Navy took all of FM's OP production for submarine use until 1944. As the Allies were gaining confidence in World War II, FM was ready to apply its highly successful OP submarine engine to the railroad market, and in 1943 the War Production Board had authorized the manufacture of a prototype 1000-h.p. FM switcher. It would be built at the new "Locomotive Department" factory on the old wagon works site in Beloit, Wisconsin. Combining established frame and truck technology from Baldwin/General Steel

Castings, their six-cylinder OP engine and Westinghouse electrical equipment, FM wrapped the entire package in a carbody styled by noted industrial designer Raymond Loewy and on August 8, 1944, rolled out their first diesel-electric locomotive: Milwaukee Road 1802, builder's number L1001. The 1802 was put to work on August 21, 1944, and survives in operable condition to this day at the Illinois Railway Museum in northern Illinois (page 128).

Four more switchers were produced using GSC cast underframes (C&NW 1036, AT&SF 500, UP DS1300 and Milwaukee 1803) before they converted over to fabricated steel underframes in August 1945.

In its early years FM had no model designations for its locomotives. However, as the model line expanded with the dual-service cab units produced at the GE shop in Erie, Pa., and 2000-h.p. "Heavy Duty" hood unit of 1947, a model designation system was devised. The switcher became the H10-44, signifying Hood carbody, 1000 horse-power and 4 motors on 4 axles (the standard "B-B" wheel arrangement). The 2000-h.p. dual service cab units never did get a model designation and are simply known as the "Erie-Builts." Some early rosters show FM "spec numbers" like Alt100.3 for the Erie passenger units; later research revealed, however, that the "Alt" stood for Advertising Literature Technical," and the spec numbers were actually technical manuals!

Between 1944 and 1963, FM would out-shop a total of 1460 locomotives with OP engines for 49 customers in the U.S., Canada and Mexico. They were occasionally ahead of their time, such as with a 2000-h.p. road switcher 13 years before EMD's GP20, and sometimes a bit late, as with the standardized "Consolidation" line of cab units. The "C-Line" of 1948 was a solid concept of horsepower and performance options (1600 to 2400 h.p. in passenger and freight versions) in modular form, unfortunately introduced in a cab unit carbody just as the railroad industry was about to be swept away by the road switcher.

But on one occasion FM was in the perfect place at the perfect time: in 1953 it introduced the 2400-h.p. Train Master, setting the stage for the high-horsepower six-motor units that would dominate the industry for decades to come. Internal corporate squabbling at the time that EMD and Alco were getting their acts together and responding to the challenge of the Train Master was FM's undoing, and FM voluntarily pulled out of the locomotive business in 1963.

THE FIRST LOCOMOTIVE POWERED by Fairbanks-Morse — with two 300-h.p. four-cycle inline engines — was the 1939 Reading centercab 97, shown alongside an EMD FT at Erie Avenue roundhouse in Philadelphia in May 1949. The 1939-built former-Southern Railway motor car 2 (**opposite**), with a five-cylinder OP engine, was at Moultrie, Georgia, on the Georgia Northern in January 1969. The first locomotive built by FM, Milwaukee Road H10-44 760 of August 1944, was working (**top**) at Sturdevant, Wisconsin, in August 1966.

WHAT'S AN OPIE?

THE SUCCESS of Fairbanks-Morse in the marine and locomotive markets can be attributed to one unique diesel engine: the two-cycle, opposed-piston model 38D8⅛. This rugged and versatile prime mover was the result of the U.S. Navy s quest for a submarine power plant in the 1930s. It has an 8⅛-inch bore and 10-inch stroke and was available in five, six, eight, ten and twelve cylinder configurations which produced 1000, 1200, 1600, 2000 and 2400 horsepower respectively.

The opposed-piston engine — universally referred to as the "OP" — is a remarkably simple machine with only one complex aspect: it has two crankshafts, one on top and one on the bottom. Within a fabricated box-like case, open-ended cylinders are lined up vertically on the centerline of the engine. Each cylinder hosts two pistons, one entering from the top and one from the bottom; each is connected to its respective crankshaft with conventional connecting rods. The two crankshafts are linked together with a vertical drive shaft at one end and are timed so that the pistons come toward each other to compress the air and then are expelled away from each other with an explosive injection of fuel at their closest point of approach (the two cranks are actually timed 12 degrees apart to provide a "lead" for exhaust scavenging, rather than being in perfect 180-degree opposition).

Using ports in the walls of each cylinder to take in fresh air and expel hot exhaust gasses, the OP engine has no cylinder heads or valves (the two-cycle EMD 567 engine has air intake ports in the cylinder walls and four exhaust valves in the cylinder heads). In the FM engine, as the bottom piston clears its side ports, the combustion gasses exit into the exhaust manifold; a split second later the top piston clears its side ports and pressurized fresh air enters, "scavenging" the cylinder and trapping a charge of fresh air for the next power stroke. As the two pistons begin to close on each other, they cover their side ports and compress the air trapped between them, driving the air s temperature up to a point where it is hot enough to ignite fuel oil. As the pistons hit their closest point of approach, fuel is sprayed into the hot air from two injectors in the side of the cylinder. The fuel instantly explodes, driving the pistons apart with great force. As the pistons clear the side ports at the opposite ends of their stroke, the cylinder scavenges to set up another power stroke.

Thus, like an EMD engine, each rotation of an OP produces a power stroke, and like the "conventional" EMD engines, the FM's are ventilated with air pressurized by a simple Roots blower, rather than a turbocharger. In service, the OP is the smoothest running engine in the field, sounding much like a very content EMD rather than a chugging and laboring Baldwin, Alco or General Electric four-cycle.

In the railroad field, the OP's strongest selling point was its simplicity and comparatively modest number of working parts (FM claimed its eight-cylinder, 16-piston 1600-h.p. OP had 237 fewer parts than EMD's 16-cylinder 567 and 118 fewer than Alco's 12-cylinder 244, both rated at 1500 h.p.). The OP's one major drawback was that the top crankshaft had to be completely removed to change out a cylinder liner, a task which could be done on an EMD 567 in less than two hours by two men with hand tools and a chain hoist. Contrary to popular myth, however, the OP's *pistons* could be changed out in a fairly simple procedure without removing either crankshaft by disconnecting the connecting rod and dropping it past the crankshaft to clear the piston from the cylinder and then removing it through side access ports.

Being a "normally aspirated" engine like the EMD, the OP was most at home — as one would suspect from a machine perfected in submarines — at sea level. The thin air of higher altitudes caused both "breathing" and cooling problems for the OP, especially in its earlier versions. The bottom crank and pistons do about 80% of the "work" in an OP, and the bottom pistons are subjected to only hot gases because of their proximity to the exhaust ports, while the top pistons are swept by the cool incoming air. Learning how to manufacture pistons that could withstand such conditions took valuable development time and hindered FM's sales and performance in its early years — this had been a minimal problem in submarines, where complex cooling systems were available.

While a properly maintained OP is a smooth and reliable engine, they tend to develop one distinctive characteristic: blue smoke on acceleration. This is the result of two potential causes, which are usually mixed to varying degrees. The first is lubricating oil. While the bottom crankshaft lives in an oil-filled crankcase (like an EMD), the top crankshaft is in a "dry sump," lubricated by

FM MODELS AND OP ENGINES

MODEL	H.P.	CYL.	R.P.M.	FIRST BUILT	DATE	LAST BUILT	DATE	TOTAL
SR Motor cars	800	5	720	SR 1	1939	AGS 41	1939	6
H10-44	1000	6	800	Milw. 1802 (760)	8/44	Indianapolis Union 18	4/50	195
Erie-Built	2000	10	850	UP 700	12/45	NYC 4405	4/49	111
H20-44	2000	10	850	UP DS1366 (FM 2000)	8/47	AC&Y 505	3/54	96
H15-44	1500	8	850	Monon 45	9/47	AC&Y 200	6/49	35
C-Line	2000	10	850	NYC 5006 (CFA20-4)	3/50	LIRR 2008 (CPA20-5)	8/50	23
H12-44	1200	6	850	Milw. 1826	5/50	CH-P 70 (301)	3/61	336
H16-44	1600	8	850	CNJ 1514	7/50	CH-P 604	2/63	299
C-Line	1600	8	850	PRR 9448A (CFA16-4)	9/50	CN 6805 (CPB 16-5)	2/55	120
H16-66	1600	8	850	C&NW 1510	1/51	Alcoa 001	1/58	59
C-Line	2400	12	850	NH 750 (CPA24-5)	4/51	NYC 4507 (CPA24-5)	3/52	22
H24-66	2400	12	850	FM Demo. TM-1	4/53	VGN 74	6/57	127
P12-42	1200	6	850	NH 3100	1/57	B&M 2	1/58	4

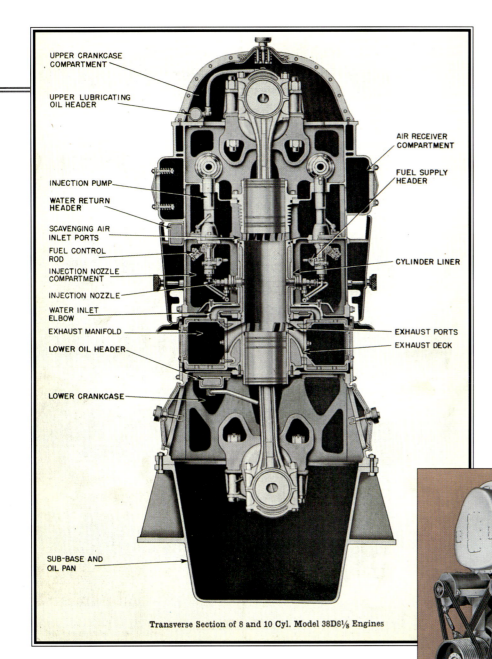

Labels on cutaway diagram:
- UPPER CRANKCASE COMPARTMENT
- UPPER LUBRICATING OIL HEADER
- INJECTION PUMP
- WATER RETURN HEADER
- SCAVENGING AIR INLET PORTS
- FUEL CONTROL ROD
- INJECTION NOZZLE COMPARTMENT
- INJECTION NOZZLE
- WATER INLET ELBOW
- EXHAUST MANIFOLD
- LOWER OIL HEADER
- LOWER CRANKCASE
- SUB-BASE AND OIL PAN
- AIR RECEIVER COMPARTMENT
- FUEL SUPPLY HEADER
- CYLINDER LINER
- EXHAUST PORTS
- EXHAUST DECK

Transverse Section of 8 and 10 Cyl. Model 38D8⅛ Engines

pressurized internal passages and a surrounding spray of oil (like the top deck valve chamber of an EMD 567). When the OP idles or shuts down, some of this top sump lube oil will drip down the cylinder walls above the piston, and if the walls are scored or the piston rings are worn, the lube oil will get into the firing chamber and often pass unburned into the exhaust manifold, where it can ignite in a smoky pall when the engine is revved up. The other cause of smoke is cooling water seeping into the cylinders from the seals where the injectors pass through the water jackets around the cylinder walls. This will also cause smoke. You could tell how well an OP is being maintained by its penchant for smoke upon acceleration.

FM's were easy to maintain by experienced mechanics, and on roads like the Virginian where they were the dominant power, they put in excellent service. On roads where they were minority units in EMD or Alco shops, however, their "different" maintenance and overhaul procedures tended to make them unwanted orphans in a hurry. Mechanics either loved 'em or hated 'em — there was little middle ground.

THE GENERATOR END of a ten-cylinder 2000-h.p. OP engine (**right**) shows the Roots blower with its air ducts to the top of the cylinder banks, while the cutaway (**above**) shows the internal components of the OP engine with the pistons at the outer ends of their strokes. In 1953 the TM-1 and TM-2 Train Master demonstrators (**below**) were working on the Duluth, Missabe & Iron Range.

The first customer

DIRECTLY SERVING the Beloit plant, the Milwaukee Road, not surprisingly, became Fairbanks-Morse's first customer for an OP-powered locomotive. Wearing builders number L1001, H10-44 1802 rolled out of the Beloit shop on August 8, 1944. Later renumbered 760, the pioneer unit was still in revenue service when it was photographed (below)

on a hot and humid August night in 1966 at Sturdevant, Wisconsin. It operates today at the Illinois Railway Museum in Union, Illinois (page 128).

From eleven 1000-h.p. switchers, the Milwaukee went for the 2000-h.p. "Erie-Built" passenger units for its newly streamlined *Olympian Hiawatha*, which made its inaugural run on June 29, 1947. Standard power for

CHICAGO, MILWAUKEE ST. PAUL & PACIFIC

MODEL	ORIGINAL NO.	RENUMBER	B/N	DATE	NOTES
H10-44	1802	760	L1001	8/44	FM's first locomotive
H10-44	1803	761	L1005	6/45	
H10-44	1804-1806	762-764	L1006-L1008	8-9/45	
H10-44	1807-1811	765-769	L1009-L1013	10-12/45	
H10-44	1812	770	L1014	8/46	
Erie A	5A, 5C	5A, 10A	L1063, L1065	10/46	
Erie B	5B	Same	L1064	10/46	
Erie A	6A-9A	Same	L1066, 69, 72, 75	1-3/47	
Erie B	6B-9B	Same	L1067, 70, 73, 76	1-3/47	
Erie A	6C-9C	11A, B; 12A, B	L1068, 71, 74, 77	1-3/47	
Erie A	22A, 22B	14A, 14B	L1102, L1104	8/47	
Erie A	21A, 21C	13A, 13B	L1119, L1120	8/47	
Erie B	21B (I)	10B	L1157	4/48	
H10-44	1819	777	10L45	3/48	Ex-FM demo, acq. 8/49
H10-44	1813-1818	771-776	10L112-10L117	1-2/49	
H10-44	1820-1825	778-783	10L325-10L330	2-3/50	
H12-44	1826, 1827	750, 751	12L374, 12L375	5/50	750 first H12-44
H12-44	1828-1832	752-756	12L428-12L432	3/51	
H12-44	1833-1837	715-719	12L448-12L452	10-11/51	
CFA16-4	23A, C-28A, C	Same	16L479-16L490	7-9/51	
CFB16-4	23B-28B	Same	16L491-16L496	7-9/51	
H12-44	1838-1847	720-729	12L520-12L529	11/51	
H12-44	2300-2309	730-739	12L561-12L570	3-4/52	
H16-66	2128-2130	553-555	16L693-16L695	9/53	Renumbered 547-549
H16-66	2125-2127	550-552	16L757-16L759	8-9/53	
H16-44	2450-2457	400-407	16L815-16L822	1/54	
H12-44	2310-2314	740-744	12L823-12L827	2/54	No cab overhang
H16-44	2463-2469	413-419	16L908-16L914	8-9/54	
H16-44	2458-2462	408-412	16L915-16L919	9/54	
H16-44	2500-2508	420-428	16L934-16L942	12/54	
H12-44	2315-2325	700-710	12L949-12L959	1-2/55	
H16-44	2509-2516	429-436	16L994-16L1001	1-2/56	

the train west of Minneapolis was an A-B-A 6000-h.p. road set, but in August 1948, it was departing Chicago (right) with only two units, led by the 6C.

In mid-1951 the Milwaukee bought six A-B-A sets of 1600-h.p. "C-Line" freight units which saw extensive use out of Chicago's Bensenville Yard south on the Chicago, Terre Haute & Southeastern to Louisville and north into Wisconsin. On October 17, 1964, two CFA16-4 cab units and an SD7 were nearing Bensenville (above) with a freight down from Milwaukee. They paused just west of O'Hare Airport to await clearance into the busy yard, presenting a handsome image. ⓜ

Pumpkins and a pink mountain

THE MILWAUKEE ROAD, at 152 units, was second only to the Pennsylvania Railroad (200 units) in ownership of FM locomotives, and nearly all except the passenger Erie-Builts survived into the 1970s. In February 1967 C-Liner 26A (above) was at Bensenville Yard as it met the brand new EMD SD45 demonstrator 4353. Sister 24A was idling the night away at Green Bay (right).

in the summer of 1964. Most Milwaukee FM's wore the orange and black "Halloween Pumpkin" colors, but H10-44s 718 and 733 were decked out in the Union Pacific-style passenger livery for use at the Milwaukee passenger station; the 718 (top) was under the old trainshed in October 1964. "Pumpkin" 763 (above right) was working Savanna, Ill., in August 1974.

The Milwaukee was one of only four customers to buy the H16-66 "Baby Train Masters;" the 551 (right) was working a Kansas City transfer job with H16-44 427 in July 1972.

Not all of the H16-44s in the condensed "Train Master" carbody were identical — note the protruding radiator section on the 405 (left) at Ladd, Ill., in June 1965. The 405 wore a handsome cast builders plate (inset). Six H16-44s were laying over at Ladd against the backdrop of one of the area's distinctive "pink mountains" of coal mine tailings. 🚂

11

Davis Junction and beyond

THE BRANCH LINE from Janesville, Wis., south through Beloit and Rockford to Ladd, Ill., crossed the Milwaukee's Chicago-Omaha main line at Davis Junction, twelve miles due south of Rockford.

In the 1960s the line was operated almost exclusively with FM H16-44s and switchers. In 1964 the 421 was idling on the east leg of the wye (left), with the tracks of the Omaha main in the foreground. On the evening of December 7, 1965, the 428 and 434 (opposite top) were holding clear of the Route 72 crossing as they made an air test in the yard before departing southward for Rochelle, Mendota and Ladd. Just north of "DJ," the Milwaukee entered CB&Q trackage rights in the yard at Rockford (below), where a southbound behind three H16-44s on May 13, 1965, passed GP7 220 next to the CB&Q roundhouse, while a C&NW SW1 could be seen through the trees. The 436 was Milwaukee's last FM unit — built in February 1956, it wears builders number 16L1001, while the Milwaukee's and FM's first unit (H10-44 760) wore builder's number L1001!

One hundred sixty miles west of Davis Junction, the 425 (opposite bottom) was working the industrial heart of Cedar Rapids, Iowa, in October 1971 in the company of EMD SW1200 633.

From Erie to Cajon

FAIRBANKS-MORSE entered the road unit market in December 1945, just 16 months after turning out its first switcher. Since the Beloit plant could not handle the nearly 65-foot carbody, the units were built at electrical-supplier General Electric's plant in Erie, Pennsylvania. The 2000-h.p. units were thus forever known simply as "Erie-Builts."

Union Pacific bought the first Erie-Builts, A-B-A 700, 700B and 701, and quickly followed with six more cabs and boosters. The Santa Fe bought its solitary A-B-A set in May 1947. The two roads' Eries soon became common sights on the joint trackage over Cajon Pass, the gateway to the Los Angeles basin. On October 8, 1950, UP 702 **(right)** had the *Los Angeles Limited* headed downhill, and Santa Fe's 90-set was dropping into San Bernardino **(opposite bottom)** on June 7, 1954. The UP 706 **(bottom)** was photographed at Summit, and the Santa Fe's 90 was at San Bernardino **(below)** in these undated views.

CARL HEHL / LOU SCHMITZ COLLECTION

K.C. HENKELS / KEITH ARDINGER COLLECTION

Yellow hoods

THE UNION PACIFIC was FM's fourth customer, buying one of the five prototype H10-44s (DS-1300) in May 1945. It returned in January 1947 for four more switchers; the 1301 was still going strong (**above**) at Council Bluffs, Iowa, on October 17, 1962 — and few paint schemes fit that Loewy carbody as well as UP's Armour yellow and gray.

In addition to the 2000-h.p. Erie-Builts and H20-44s, the UP bought five boiler-equipped H15-44s for local passenger and mixed train service on prairie branch lines. The 1329 (**left**) was at Council Bluffs in October 1964. The three H16-44s of 1950 also had the Loewy-style carbody. 🔄

UNION PACIFIC — SOUTHWEST PORTLAND CEMENT

MODEL	NUMBER	1947 RE#	1948 RE#	1955 RE#	B/N	DATE	NOTES
H10-44	DS1300	—	—	1300	L1004	5/45	
H10-44	DS1301-DS1304	—	—	1301-1304	L1025-L1028	1-2/47	
H20-44	DS1365, DS1366	—	—	1365, 1366	L1031, L1032	8, 10/47	1366 FM Demo. 2000. To SWPC 66 in 1963 (re# 409)
H20-44	DS1360-DS1364	—	—	1360-1364	L1033-L1037	8/47	
H20-44	DS1367-DS1370	—	—	1367-1370	L1040-L1043	11-12/47	1369 to SWPC 69 in 1963 (re# 408)
Erie A	50M-1-A	981A	700	650	L1060	12/45	
Erie A	50M-2-A	982A	701	651	L1061	12/45	
Erie B	50M-3-B	983B	700B	650B	L1063	12/45	
Erie A	984A, 985A	—	702, 703	652, 653	L1117, L1118	6/47	
Erie B	986B, 987B	—	702B, 703B	652B, 653B	L1127, L1128	10/47	
Erie A	704-707	—	—	654-657	L1136-L1139	3-4/48	
Erie B	704B, 706B	—	—	654B, 656B	L1142, L1143	3, 4/48	
H15-44	DS1325-DS1329	—	—	1325-1329	15L7-15L11	4-5/48	All boiler equipped
H16-44	DS1340-DS1342	—	—	1340-1342	16L370-16L372	8/50	1340 later boiler equipped
H20-44	AC&Y 505	—	—	—	20L832	3/54	To SWPC 410 in 1971

NOTE: Previously published rosters show SWPC 408 as being ex-UP 1366, ex-FM demonstrator 2000; Illinois Railway Museum discovered that their 409 (**page 128**) is actually the 1366, not 1369, as numerous internal parts are marked "1366." The 408 is at the San Diego Railroad

... and Victorville's second-handers

PROBABLY THE MOST DISTINCTIVE FM's on the UP roster were its eleven 2000-h.p. H20-44s, which were used extensively as helpers on Cajon Pass. In 1963 two were sold to the Southwest Portland Cement Company to replace GE 70-tonners on its 15-mile line out of Victorville, Cal., to a limestone quarry in the desert (right), where SWPC 408 (ex-UP 1369) was photographed in May 1976. In 1971 SWPC added Akron, Canton & Youngstown 505, which became SWPC 410, shown at Victorville (top and below) in October 1971. All three are now in museums.

Did you say 2000 horsepower?

<table>
<tr><td colspan="5">Akron, Canton & Youngstown</td></tr>
<tr><th>Model</th><th>Numbers</th><th>Date</th><th>B/N</th><th>Notes</th></tr>
<tr><td>H20-44</td><td>500-503</td><td>1-2/48</td><td>L1044-L1047</td><td></td></tr>
<tr><td>H15-44</td><td>200</td><td>6/49</td><td>15L36</td><td></td></tr>
<tr><td>H20-44</td><td>504</td><td>2/51</td><td>21L309</td><td></td></tr>
<tr><td>H16-44</td><td>201-203</td><td>12/51</td><td>16L530-16L532</td><td></td></tr>
<tr><td>H16-44</td><td>204, 205</td><td>4/54</td><td>16L783, 16L784</td><td></td></tr>
<tr><td>H16-44</td><td>206</td><td>3/54</td><td>16L831</td><td></td></tr>
<tr><td>H20-44</td><td>505</td><td>3/54</td><td>20L832</td><td>Last H20-44 built. To SWPC 410 in 1971</td></tr>
<tr><td>H16-44</td><td>207</td><td>12/55</td><td>16L993</td><td></td></tr>
<tr><td>H16-44</td><td>208</td><td>3/57</td><td>16L1156</td><td></td></tr>
<tr><td>H20-44</td><td>506-508</td><td>2/53</td><td>21L714-21L716</td><td>P&WV 68, 69, 67 to AC&Y 1967-'68</td></tr>
</table>

THE FIRST H20-44 I ever saw was on a trip to the East in August 1966 when I encountered Akron, Canton & Youngstown 500 (opposite bottom) in the engine terminal at Delphos, Ohio. I found it hard to believe that this "switcher" packed 2000 horsepower. But I later learned that it was FM's first venture in building a true road freight unit. After creating the monster Erie-Built "dual-service" cab units — which with A1A trucks were really passenger units — in 1947 FM packed the same ten-cylinder OP engine into an end-cab carbody mounted on B-B road trucks to produce what it termed its "Heavy Duty" freight locomotive, the most powerful unit of its time. With competing freight units pegged at 1500 h.p., FM almost squandered its horsepower advantage by pitching the Heavy Duty as a low speed drag engine. It would be a dozen years before the concept of a 2000-h.p.

high speed road unit would sweep the railroad industry in the form of EMD's turbocharged GP20.

The 169-mile AC&Y embraced the Heavy Haul (later designated H20-44), buying six of them, including the last built, AC&Y 505. It also bought one H15-44, shown (left) at Akron on May 27, 1961, and eight H16-44s (the AC&Y also rostered seven Alco switchers and one RS1 road switcher). The AC&Y was taken over by the expanding Norfolk & Western in 1964 but operated as a separate entity, and in 1967-'68 it picked up three more H20-44s from the N&W's Pittsburgh & West Virginia.

Meanwhile, back in Ohio in August 1966, somewhere around Bluffton we spotted a westbound headlight on the AC&Y which turned out to be H20-44 501 (above and opposite) rambling through the cornfields en route to its interchange with the Nickel Plate (now also N&W) at Delphos. A 2000-h.p. road unit a decade ahead of its time? It just looked like an overgrown switcher to me. ◑

Destination: Mogadore

TWO YEARS AFTER my encounters with the H20-44s on the west end, in October 1968 I returned to the AC&Y at its yard in Akron, Ohio. In the engine terminal I found H16-44 207 (below) idling on the turntable in the company of H20-44 507 (bottom), former Pittsburgh & West Virginia 69. In the yard, the 205 was about to depart for the east end of the railroad: the distinctively named "Mogadore." It turned out to be an area of light industry where the 205 went to work shuffling boxcars for local distribution (right). It put on a fine show, shoving up and down the lead beneath occasional appropriate bursts of blue exhaust smoke.

I'd been told that there was a work train out on the line, and as the late afternoon sun was dropping low I found H16-44 208 and H20-44 508 (ex-P&WV 67) handling a wrecking crane that was picking up derailed freight cars — the 508 was the culprit producing the picturesque exhaust plume (opposite bottom). It was the distinctive signature of an FM. ⟲

![Rock Island logo]

Re-engined Rockets

MATT HERSON

MEETING THE COMPETITION of 1947 with a 1500-h.p. road switcher, FM introduced the "All Purpose" H15-44 in a distinctively curved high-hood carbody styled by Raymond Loewy. The Chicago, Rock Island & Pacific bought two of them in December 1948 specially equipped for Chicago suburban service with steam heat boilers and provisions for head-end coach lighting power. The Rock's only FM locomotives, in May 1958 the 400 and 401 were re-engined by EMD at La Grange, Ill., with 1500-h.p. 16-cylinder 567C engines, creating essentially GP7s in FM carbodies. In July 1961 the 401 was back in suburban duty (top) with a southbound train at Englewood, and on Halloween Day 1964 (left), the 400 was idling alongside EMD *Aerotrain* No.3 at the 49th Street passenger diesel shop. In June 1965, the pair was in freight service (bottom) at Joliet behind an F7.

CHICAGO, ROCK ISLAND & PACIFIC			
MODEL	NUMBERS	B/N	DATE
H15-44	400, 401	15L12, 15L13	12/48

John Barriger's Babies

CHICAGO, INDIANAPOLIS & LOUISVILLE			
MODEL	NUMBER	B/N	DATE
H10-44	18	L1020	11/46
H15-44	45, 46	L1198, L1199	9, 12/48

JOHN WALKER BARRIGER III was just beginning to make a name for himself in 1945 when he left Fairbanks-Morse as manager of its diesel engine and railroad division to become president of the Chicago, Indianapolis & Louisville — better known as the "Monon." Dieselizing the Monon with EMD F-units and Alco RS2s, Barriger bought only three FM's to his new home: H10-44 No.18 (right, at East Chicago, Ind., on July 6, 1964) which worked nearly its entire career as the Lafayette yard engine, and H15-44s 45 and 46, which were re-engined by EMD in 1960. On December 21, 1964, the 45 and an F3B (top) were handling a transfer at 73rd Street in Chicago, while the 46 posed at Hammond, Indiana, (below) in 1966.

DR. ART PETERSON COLLECTION

 **An H10-44 or an H12-44?**

YOU CANT IDENTIFY the model of an FM switcher by looking at its carbody. When FM upgraded its 1000-h.p. switcher to 1200 h.p. by increasing the r.p.m.'s on the six-cylinder engine from 800 to 850, it retained the original Loewy-styled carbody with the distinctive cab roof overhang. Minneapolis, Northfield & Southern 11 (**opposite top**, at Minneapolis in November 1967) was an H10-44, built in 1946 as Minnesota Western 51, while MN&S 10 (**opposite bottom**, same place and date) was a 1951-built H12-44 — but the two appear to be identical. In contrast, neighbor Soo Line bought its five H12-44s in two orders: 315 in June 1952 and 316-319 in December 1954. In September 1952 FM changed to the more spartan squared-off carbody for the H12-44; thus, Soo Line 315 (**below**, on July 28, 1965, in the original black and gold livery and **below right** in August 1966, both in Minneapolis) had the classic Loewy carbody, while the others like the 317 (**above**, at Northtown Yard in Minneapolis in July 1973) were delivered in the spartan carbody. The last switcher in the

Loewy body was Santa Fe 530, built September 17, 1952, and the first spartan carbody was Southern Pacific 1486 (**page 102**).

MINNEAPOLIS, NORTHFIELD & SOUTHERN				
MODEL	B/N	DATE	MW No.	MN&S No.
H10-44	L1019	9/46	51	11
H12-44	12L427	1/51	10	10

SOO LINE			
Model	B/N	Date	Number
H12-44	12L636	6/52	315
H12-44	12L965-12L968	12/54	316-319

Chicago — north and west

SINCE IT ALSO SERVED Beloit, Wisconsin, the Chicago & North Western was the logical second customer for FM locomotives, and in November 1944 it took delivery of FM's second "prototype" H10-44, C&NW 1036, b/n L1002. The first five switchers (Milwaukee 1802, C&NW 1036, AT&SF 500, UP DS1300 and Milwaukee 1803) were built on General Steel Castings underframes designed for Baldwin switchers. The C&NW bought six more production H10-44s before acquiring four Erie-Built passenger units for its subsidiary Chicago, St. Paul, Minneapolis & Omaha in the spring of 1947. "Omaha Road" 6001B (opposite center) and a companion were at the 40th Street passenger engine terminal in Chicago in 1948. The big Eries were not restricted to Omaha Road trains and were fairly common on the Chicago-Milwaukee-Minneapolis main line runs and the *Peninsula 400* to Green Bay and Ishpeming.

The C&NW's FM switchers (26 H10-44s and nine H12-44s) were used all across the system. In June 1967 (opposite bottom) an H10-44 was adding a high-level diner, coach and baggage/lounge and deadhead coach to the rear of the Chicago-bound *Flambeau 400* at Green Bay while two E8s led by 5022A backed onto the Ashland-bound *Flambeau 400*. H10-44 1064 (opposite top) was at the Proviso engine shop in west suburban Chicago in 1966.

When it comes to FM power, however, the C&NW was best known for its fleet of H16-66s. While they gained lasting notoriety for surviving into the 1970s in the iron ore country above Green Bay, in

the 1960s they were often seen around Chicago, Milwaukee, Janesville and Beloit. In December 1965 (above) four H16-66s were lined up on the ready track at Proviso in the company of GP35 858. They would likely soon be heading up tonnage for Milwaukee and Fond du Lac. ⊕

CHICAGO & NORTH WESTERN

Model	Number	Date	B/N
H10-44	1036	11/44	L1002
H10-44	1048-1051	19-10/46	L1015-L1018
H10-44	1052, 1053	11/47	L1085, L1086
Erie A	6001A, 6001B (CStPM&O)	6/47	L1090, L1092
Erie A	6002A, 6002B (CStPM&O)	5/47	L1093, L1095
H10-44	94 (CStPM&O)	12/47	L1171
H10-44	1054, 1055	1/48	L1172, L1173
H10-44	95-98 (CStPM&O)	2/48	10L40-10L43
H10-44	1082	6/48	10L59
H10-44	1056-1062	6-7/49	10L134-10L140
H10-44	1063-1065	12/49-1/50	10L284-10L286
H10-44	1070	1/50	10L287
H12-44	1071, 1072	5/50	12L376, 12L377
H16-66	1510, 1511 (Loewy carbody)	1/51	16L35, 16L37
H16-66	1512-1514	2/51	16L275-16L277
H16-66	150 (CStPM&O)	2/51	16L278
H16-66	1605-1612	10-11/52	16L659-16L666
H12-44	1110-1113	12/52	12L709-12L712
H16-66	1668	7/53	16L696
H16-66	1669-1673	7/53	16L699-16L703
H16-66	168-172 (CStPM&O)	7/53	16L704-16L708
H12-44	1114-1116	8/53	12L768-12L770
H16-66	1674-1683 ("Baby Train Master")	7-8/54	16L872-16L881
H16-66	1696-1698	8-9/55	16L972-16L974
H16-66	1694, 1695	11/55, 9/55	16L981, 16L982
H16-66	1691-1693	9, 10/55	16L983-16L985
H16-66	1700, 1699	9/55	16L986, 16L987
H16-66	1901-1903	6/56	16L1003-16L1005
H16-66	1904-1906	6/56	16L1029-16L1031

The unique Loewy H16-66s

B Y 1950, THE BALDWIN Locomotive Works had been successfully marketing a 1500-h.p. six-motor road switcher for two years, and Alco was tooling up for a similar unit of its own. Detecting a potential market, FM put its H16-44 power package (eight-cylinder OP engine and Westinghouse electrical equipment) on an elongated frame and added the proven Baldwin 13-foot rigid-bolster three-motor truck to produce its first six-motor locomotive. The rounded Loewy carbody in use on the H16-44s of that time was retained. The first H16-66 was C&NW 1510 (below, at Escanaba in August 1966), delivered in January 1951. It was followed by

24 more similar units delivered through July 1953. Six of the Loewy H16-66s were assigned to the subsidiary CStPM&O and carried three-digit numbers. In August 1972, Omaha Road 150 (bottom) was idling outside the Green Bay roundhouse, while sister 170 (right) was riding the Escanaba turntable the next day, about to be put into the roundhouse. Back in December 1968, Omaha 169 and C&NW 1610 (opposite bottom) were stored dead for the winter outside the Green Bay roundhouse alongside the 1677, one of the newer H16-66s in the "Baby Train Master" carbody introduced on the C&NW deliveries of mid-1954.

KERMIT GEARY, SR.

28

FAIRBANKS-MORSE unveiled its massive 2400-h.p. six-motor Train Master at the Association of American Railroads convention in Atlantic City, N.J., in June 1953. Very pleased with the 25 Loewy H16-66s it already had in service, the C&NW was on the books for additional H16-66s. Taking advantage of some of the technology developed for the Train Master, FM redesigned the carbody of the H16-66 to resemble that of the Train Master and placed it on the big unit's new Tri-Mount drop-equalizer trucks. The resulting units, the first six of which were delivered to the Milwaukee Road in August 1953, were immediately dubbed "Baby Train Masters." The C&NW got its first Baby Train

Masters in July 1954. With this change in styling, the C&NW became the only railroad to own the Loewy-styled H16-66s. While the six Milwaukee H16-66s retained the Westinghouse electrical equipment, the C&NW's new units were equipped with General Electric motors and systems, and unlike the Loewys, which were long hood forward, the Baby Train Masters were set up short hood forward.

By mid-1956 the C&NW had amassed a fleet of 51 H16-66s (25 Loewys and 26 Baby Train Masters), which by the late 1960s had found a home in the iron ore country of Wisconsin and upper Michigan. In August 1972 the 1695, 1696 and 1905 (above) were getting an ore train east

JEREMY PLANT

out of Negaunee, while Milwaukee Road H16-66 549 (**above**) was at the Escanaba roundhouse for pool service between there and Iron Mountain. The 1682 and 1696 were at Fond du Lac in July 1965 (**left**). The first C&NW Baby, 1674, was on the Fond du Lac turntable in December 1968 (**below**) while the last C&NW H16-66, 1906, idled alongside. In July 1973 the 1904, 1699 and a third H16-66 (**opposite bottom**) had an ore train eastbound at Waucedah, Wisconsin. The FM's were replaced by ex-N&W Alco C628s in early 1975.

508
at
Janesville

A MAIL AND EXPRESS CAR was loaded and ready to be added to the end of C&NW Train 508 at Janesville, Wisconsin, on a Sunday evening in September 1965 by H12-44 switcher 1110. The boxy FM idled contentedly in front of the station (left and right) about 20 minutes before the 8:30 arrival of the Sunday-only remnant of the old *Viking* from Madison. An airhorn to the north announced the approach of 508, and the FM moved up to be in position for work as soon as the E7 (center right) eased to a stop. With the mail car tacked on and the kerosene markers in place (below), 508's headlight kicked up to bright for departure. As 508 disappeared toward Chicago and the FM went back to the yard, a weatherbeaten sign on the depot (far right) was a quiet reminder of more prosperous times in Janesville.

"A rattler on the Haywire"

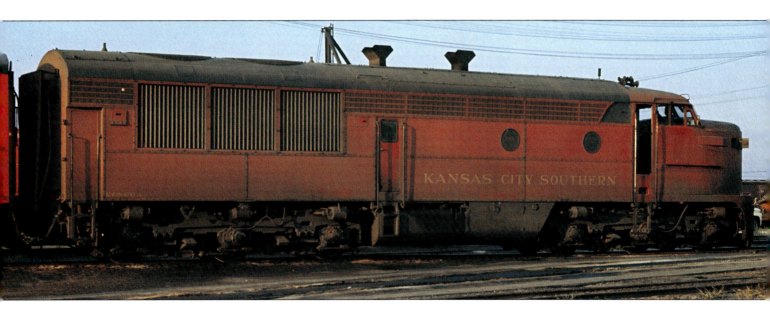

THE PHONE RANG in Sheffield Tower just east of Union Station in Kansas City. "Ya' got a rattler on the Haywire," came the message from the next operator to the south. It was an August morning in 1965, and a "rattler on the Haywire" translated to a "freight on the Kansas City Southern." A few minutes later our "rattler" came into view with a whole mish-mash of "covered wagons" behind Erie-Built 61A: an F7B, an Erie B, another F7B and Erie cab unit 60A. In spite of appearances, the consist that banged across the Santa Fe diamonds was 100% EMD-powered, because in the mid-1950s the KCS had EMD re-engined its three A-B-A Erie-Built sets with 1750-h.p. 16-cylinder 567C's. The same thing then occurred in 1958 when one of the KCS' two H15-44s got the same treatment; KCS 41 was renumbered 45 shortly thereafter (actually, one Erie set, 62A-B-C and the 45 were carried on the roster of KCS-subsidiary Louisiana & Arkansas). The 45, in fresh white paint (opposite bottom) was at

Heavener, Okla., in May 1968, while the 40, which was never re-engined, was at Pittsburg, Kan., on January 3, 1963 (opposite center).

After crossing the Santa Fe, our "rattler" tied up in the "Joint Agency" Milwaukee Road/KCS engine terminal (above) alongside a Milwaukee U25B and Alco S4 switcher. The 61A's builders plate (right) identifies GE's Erie Works. ⊕

KANSAS CITY SOUTHERN

MODEL	B/N	DATE	NUMBERS
Erie A	L1087, L1089	11/46	60A, 61A
Erie B	L1088, L1091	11/46	60B, 61B
Erie A	L1094, L1096	1/47	60C, 61C
Erie A	L1098, L1097	6/47	L&A 62A, 62C
Erie B	L1145	4/48	L&A 62B
H15-44	15L28, 15L29	5/49	L&A 40, 41 (41 re# 45 4/58)

34

FAIRBANKS, MORSE & CO.
CHICAGO, ILLINOIS, U.S.A.
DIESEL ELECTRIC LOCOMOTIVE
SERIAL NO.
L-1089   GE28584
DATE
NOV. 1946
BUILT AT ERIE, PA. WORKS — GENERAL ELECTRIC CO.

W. WOELFER COLLECTION

JOE VAN HOOREBEKE

 Three Midwestern Terminals

THE SMOOTH RHYTHM of FM's OP engine could be heard around many Midwestern cities. The Terminal Railroad Association of St. Louis sampled just about everything in switchers, including four H10-44s; "Trah-Rah" 700 (left) was working the west leg wye of St. Louis Union Station in June 1965.

The Kentucky & Indiana Terminal of Louisville, Kentucky, had ten H10-44s and seven H12-44s; the 65 displayed its impressive new 1960s livery (opposite top) at Youngtown Yard in September 1965, while the 63 (below) in August 1971 was in simplified blue. H12-44s 60-64 were delivered in Loewy carbodies but later had their cab overhangs trimmed off.

The Indianapolis Union Railway had nine H10-44s (like the 15, above, on July 21, 1959) and three H12-44s, which were among the first in the spartan carbody.

TERMINAL R.R. ASSN. OF ST. LOUIS			
Model	Number	Date	B/N
H10-44	701, 700	3/47	L1029, L1030
H10-44	702, 703	4/49	10L118, 10L119
KENTUCKY & INDIANA TERMINAL			
H10-44	48-50	10-11/47	L1082-L1084
H10-44	51, 52	3/48	10L44, 10L46
H10-44	55-59	5/49	10L141-10L145
H12-44	60-64	3-4/51	12L402-12L406
H12-44	65, 66	3/53	12L719, 12L720
INDIANAPOLIS UNION RY.			
H10-44	10-13	8, 11/49	10L300-10L303
H10-44	14-18	3-4/50	10L363-10L367
H12-44	19-21	9-10/52	12L650-12L652

Big blue hoods

THE SANTA FE was the third customer to buy a Fairbanks-Morse locomotive: H10-44 500 in April 1945, one of the three prototypes with cast Baldwin underframes. Two years later it bought an A-B-A set of Erie-Builts (page 44) and returned the following year for another pair of H10-44s. While the Erie-Builts were plagued with problems, the switchers proved reliable and economical, and in the early 1950s the Santa Fe ordered 28 of the upgraded 1200-h.p. switchers, which retained the Loewy carbodies. In August 1963 (opposite

bottom) H12-44 510 was in its as-delivered black "switcher" scheme, while 513 was in the newer blue and yellow hood unit livery. The 20 H16-44s were also delivered in the black with silver stripes, but the two orders had one visible difference. The first ten, built in 1951, had the classic Loewy rounded cab windows, evident on 3001 (right) at Argentine, Kan., alongside E6 14 in August 1966, while the 1952 group had conventional square windows.

Between 1953 and 1957 the Santa Fe returned to FM for 31 more H12-44s in the spartan carbody, typified by the

554 (below), which was at Emporia, Kan., in May 1972, and the 557 (top left), which was busy shoving Amtrak's *Southwest Chief* into Chicago's Union Station in August 1973.

DON WALLWORTH

ATCHISON, TOPEKA & SANTA FE

MODEL	NUMBER	B/N	DATE	NOTES
H10-44	500	L1003	4/45	On cast Baldwin frame
Erie A-B-A	90L, 90A, 90B	L1099-L1101	5/47	
H10-44	501, 502	10L56, 10L57	4/48	
H12-44	503-506	12L388-12L391	9-10/50	Loewy carbody
H12-44	507-516	12L438-12L447	10-11/51	Loewy carbody
H16-44	2800-2809	16L501-16L510	3-6/51	Re# 3000-3009
H16-44	2810-2819	16L589-16L598	6-7/52	Re# 3010-3019
H12-44	517-530	12L599-12L612	7-9/52	Loewy carbody
H12-44	531-540	12L747-12L756	4/53	Spartan carbody
H12-44	544-558	12L1006-12L1020	12/55-4/56	Spartan carbody
H12-44TS	541-543	12L1021-12L1023	5/56	Steam boiler equipped
H12-44	559-564	12L1094-12L1099	3-4/57	Spartan carbody

Chicago's steam switchers

SHUFFLING PASSENGER CONSISTS between Chicago's Dearborn Station and its 21st Street Coach Yard was a problem for the Santa Fe because the trains had to be supplied with steam heat during the moves. To solve the problem, in May 1956 Santa Fe had FM build three unique boiler-equipped switchers. These "H12-44TS" units were essentially H12-44s placed on elongated underframes with the steam generator placed in a short hood behind the cab. Riding on switcher trucks, in spite of appearances, these units were not road switchers. In January 1966 (above) all three were visible behind Alco PA 54 at the 18th Street roundhouse. The 542 (opposite) was idling clear of the about-to-depart *Super Chief* in April 1970, while the 541 was venting steam in May 1966. The 543 is preserved today (right) at the California State Railroad Museum in Sacramento.

40

Chicago switchers

GARBED IN BLACK like the set-dressers in Japanese Kabuki Theater, switchers scurried about the Chicagoland stage almost unnoticed among the "stars" of the high iron. On July 10, 1966, Pennsylvania Railroad H10-44 9098 (**right**) was at 21st Street Junction turning the consist of No.95-75, the combined *Buckeye* and *Kentuckian* while an A-B-B-B-A set of Santa Fe warbonnets was about to clatter across the diamonds on its way into Dearborn Station with No.20, the *Chief.*

Back on June 16, 1964, Pennsy H10-44 9297 (**below**) had a caboose-hop northbound at Western Avenue Junction on the old Panhandle route into Union Station while Wabash GP9 491 was headed back to Landers Yard with a transfer job.

Of course, Chicago wasn't the only place where the Pennsy used its FM switchers; witness the undated portrait of the 9198 (**opposite**) taken at Walbridge Yard near Toledo, Ohio. 🐾

Pennsylvania Railroad

Model	B/N	Date	PRR Numbers
Erie B	L1103, 6, 9, 12, 15	11-12/47	9456B-9464B (even only))
Erie A	L1108, 10, 11	11-12/47	9456A-9458A
Erie A	L1113, 14, 16	12/47-1/48	9459A-9461A
Erie A	L1121-L1124	1-2/48	9462A-9465A
Erie B	L1125, L1126	12/47	9466B, 9468B
Erie B	L1129	12/47	9470B
Erie A	L1130-L1135	2-3/48	9466A-9471A
Erie A	L1140, L1141	6/48	9472A, 9473A
Erie B	L1144	7/48	9476B
Erie B	L1146, L1147	6/48	9472B, 9474B
Erie A	L1148-L1153	6-9/48	9474A-9479A
Erie A	L1154-L1156	10/48	9480A-9482A
Erie B	L1158	9/48	9478B
Erie A	L1161-L1169	10-12/48	9483A-9491A
H20-44	20L37-20L41	12/48-1/49	9305-9309 (PC 7705-7709)
H20-44	20L42-20L46	1-2/49	9300-9304 (PC 7700-7704)
H20-44	20L47, 20L48	2-3/49	9310, 9311 (PC 7710, 7711)
H10-44	10L61-10L63	6/48	5980-5982 (PC 8211-8213)
H10-44	10L66-10L69	6/48	5983-5986 (PC 8214-8217)
H10-44	10L70-10L82	6-9/48	9184-9196 (PC 8218-8230)
H10-44	10L96-10L98	9/48	5997-5999 (PRR 9197-9199 / PC 8231, 8232 [9199 not renumbered])
H10-44	10L120-10L131	3-4/49	9288-9299 (PC 8233-8243 [9297 not renumbered])
H10-44	10L150-10L169	7-11/49	9080-9099 (PC 8244-8262 [9088 not renumbered])
CFA16-4	16L339-16L346	9-10-50	9448A-9455A
CFA16-4	16L347-16L354	10-11/50	9492A-9499A
CFB16-4	16L355-16L358	9-10/50	9448B-9454B (even only)
CFB16-4	16L359-16L362	10-11/50	9492B-9498B (even only)
H20-44	21L453-21L478	4-6/51	8917-8942 (PC 7717-7742)
H16-44	16L579-16L588	4-5/52	8807-8816 (PC 5150-5159)
H12-44	12L637-12L649	11/52	8711-8723 (PC 8327-8339)
H12-44	12L828-12L830	2/54	8708-8710 (PC 8340-8342)
H24-66	24L897	8/56	8699 (PC 6708)
H24-66	24L898-24L905	8-9/56	8700-8707 (PC 6700-6707)

Hump Day

ALTHOUGH IT DIDN'T buy its first FM locomotives until the end of 1947, the Pennsylvania Railroad went on to become far and away Beloit's biggest customer, acquiring 200 units ranging from H10-44s to Train Masters. On February 1, 1968, nearly all of the Pennsy's FM's were joined with the New York Central fleet under the corporate banner of Penn Central, which immediately began to thin the ranks of its OP's and tried to concentrate its remaining FM's in the greater Chicago area.

One day in September 1968, PC H16-44 5158 and Train Master 6700 (former Pennsy 8815 and 8700) were working the 59th Street Yard in Chicago (**right** and **bottom**), far from their more typical haunts like Pitcairn Yard on the east side of Pittsburgh (**below**), where H16-44 8813 was working in September 1964, or Conemaugh Yard (**opposite bottom**) in Johnstown, Pa., where Train Masters worked as helpers toward Altoona. On August 16, 1958, H20-44s 9305 and 9306 were idling at Conemaugh alongside three huge Baldwin center-cab transfer units.

EMERY GULASH

Brunswick green freighters

THE BIG 2000-H.P. ERIE-BUILTS of 1945 were immediately attractive to the Pennsylvania Railroad, which was still developing its own duplex steam locomotives in that year. In late 1947 the Pennsy began buying the big Eries, and within only one year they had accumulated a fleet of 48 units (36 cabs and a dozen boosters), which would turn out to be 43% of FM's total production! Although configured as passenger units with A1A trucks, the Pennsy regarded its Eries as freight units and painted them in the single-stripe Brunswick green freight livery, although twelve of them were equipped with boilers for passenger service. In December 1956 (below) the 9470 and a mate had

a coal train westbound at Marysville, Pa, and the 9471 (above right) was at Detroit on May 16, 1961. The Eries were built with cooling systems based on U.S. Navy submarine standards; the massive size of the radiators are evident even from the outside.

When the C-Line freight units were introduced in 1950 with production in Beloit, the Pennsy opted for eight A-B-A sets of CFA16-4 1600-h.p. units. The 9496 (above left) was at Jeffersonville, Indiana, in August 1964, while the 9492 (right) was heading up a CFB16-4 and flock of Alcos at East Altoona on September 8, 1959. In sparkling fresh paint, the 9494 (opposite bottom) was leading a light engine move at Akron, Ohio, in July 1963.

MATT HERSON

N EW YORK CENTRAL subsidiary Pittsburgh & Lake Erie got the system's first FM power in the form of two H10-44s in December 1946 (page 50), but the parent company followed up in less than a year with a pair of Erie-Builts for freight service, followed quickly by four more freight cabs and a pair of boosters in the two-tone gray lightning stripe livery. Six passenger cabs — with silver painted trucks — came along in the spring of 1949. Like its rival Pennsy, the NYC continued buying FM "covered wagons" in the form of freight C-Liners, opting initially, however, for a dozen 2000-h.p. CFA20-4 cab units (5006-5017) and three boosters (5102-5104). Eight CFA16-4 1600-h.p. cabs (6600-6607) and four boosters (6900-6903) in 1952 completed the freight fleet, while eight 2400-h.p. CPA24-5s (4500-4507) were acquired for passenger service. In the late 1950s, most of the Eries and all of the 2000-h.p. C-Liners were re-engined with 16-cylinder EMD 567C's.

Re-engined Erie 5003 (above) was photographed at night in West Detroit in January 1963, while passenger Erie 4401 and CFA24-5 4506 (below) were at North White Plains, N.Y., in August 1960. The re-engined CFA20-4 5011 (opposite top) was photographed in February 1962.

The New York Central acquired a respectable fleet of FM switchers (seven H10-44s and 27 H12-44s), all in the Loewy carbody. On October 11, 1967, Penn Central 8305, 8325 and 8318 (right) were teamed up at Frontier Yard in Buffalo. In 1970, H12-44s 8309 and 8310 were sold to U.S. Steel's Fairless Works (page 126).

MATT HERSON

NEW YORK
CENTRAL
SYSTEM

MATT HERSON

ALLAN H. ROBERTS

New York Central System

Model	Railroad	Number	B/N	Date	Notes
H10-44	P&LE	9100, 9101	L1023, L1024	12/46	PC 8200, 8201
Erie A	NYC	5000, 5001	L1105, L1107	10/47	Freight; 5001 re-engined EMD
H20-44	IHB	7110-7114	20L18-20L22	7/48	
H20-44	NYC	7100-7109	20L23-20L32	7-10/48	
H10-44	P&LE	9102, 9103	10L64, 10L65	8/48	PC 9102, 9103
H20-44	IHB	7115, 7116	20L35, 20L36	9/48	
Erie A	NYC	5002	L1170	12/48	Freight; re-engined EMD
Erie B	NYC	5100, 5101	L1159-L1160	12/48-1/49	Freight; 5101 re-engined EMD
Erie A	NYC	5003-5005	L1174-L1176	1-2/49	Freight; 5003 re-engined EMD
Erie A	NYC	4400-4405	L1177-L1182	3-4/49	Passenger; all re-engined EMD
H20-44	IHB	7117, 7118	20L49, 20L50	4, 5/49	
H10-44	NYC	9104, 9105	10L146, 10L147	6/49	PC 8204, 8205
H10-44	NYC	9106-9109	10L170-10L173	11/49-1/50	PC 8206-8209
H10-44	PC&Y	1	10L175	6/49	
H10-44	NYC	9110	10L176	2/50	PC 8210
CFB20-4	NYC	5102-5104	21L270-21L272	5-7/50	
CFA20-4	NYC	5006-5017	21L288-21L299	3-7/50	Re-engined EMD 1955-56
H12-44	NYC	9111-9120	12L378-12L387	11/50-1/51	9114-9120 to PC 8303-8309
H16-44	NYC	7000-7012	16L414-16L426	7, 10/51	PC 5100-5112
CFA16-4	NYC	6600-6607	16L541-16L548	2/52	
CFB16-4	NYC	6900-6902	16L549-16L551	2/52	
CPA24-5	NYC	4500-4507	24L552-24L559	3/52	
CFB16-4	NYC	6903	16L560	2/52	
H12-44	NYC	9121-9137	12l613-12L629	5-6/52	PC 8310-8326

A SIMPLE BLACK LIVERY with minimal striping adorned the New York Central System switchers. The company's first FM's were two H10-44s in December 1946 for subsidiary Pittsburgh & Lake Erie, which got another pair two years later. The first and last P&LE switchers, 9100 and 9103 (bottom left), were at McKees Rocks, Pa., on February 11, 1961. The affiliated Pittsburgh, Chartiers & Youghiogheny got its solitary FM, No.1, (left, on Neville Island on April 17, 1966) in June 1948.

The Central's road units, however, were given the lightning stripe livery, which had some variations. The Wallkill Valley branch job down from Kingston (above) was at Campbell Hall, N.Y., on St. Patrick's Day 1957 with two of the 13 H16-44s bought in 1961. The 7002 wears the lightning stripe on a black carbody, while trailing 7003 has the same paint scheme on a gray carbody. The NYC used five miles of trackage rights over the Erie from Montgomery to reach Campbell Hall, where it had interchanges with the Lehigh & New England and New York, Ontario & Western.

The gray lightning stripe livery was also later applied to the New York Central's H20-44s. The 7108 (opposite top), shown at East Alton, Ill., on October 16, 1964, had been delivered in 1948 in black with the gray lightning stripe ending in front of the cab in

DON WALLWORTH   RICHARD R. WALLIN / MATT HERSON COLLECTION

ALAN MILLER COLLECTION

the same pattern as Indiana Harbor Belt 7112 (above). Given class designation "DFT-1," the H20-44s were regarded by NYC as freight transfer units. The FM "Heavy Duty" road units found an immediate home replacing huge three-cylinder 0-8-0s on the Indiana Harbor Belt, which handled nearly all the NYC's interchange traffic within the greater Chicago area, looping west from Gary, Indiana, and Gibson Yard in Hammond around the city to the C&NW, Soo Line and Milwaukee Road on the northwest side. The IHB units, such as the 7112, shown at Blue Island, Ill., in this undated photo, were delivered in "Pacemaker" green with an orange stripe and red trim. Note the distinctive headlight visor. The H20-44s had a short career on the IHB, as they were replaced by EMD switchers and transferred to the NYC in 1948 (7115 and 7116) and 1951 (7110-7114 and 7117 and 7118). ✦

THOMAS J. McNAMARA

PHILLIP C. FAUDI / W. WOELFER COLLECTION

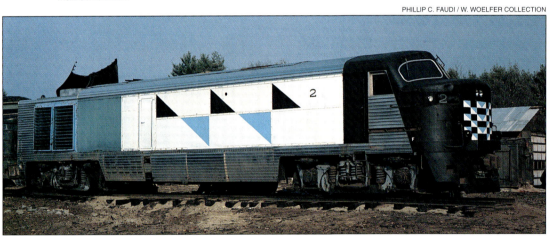

THOMAS J. McNAMARA

PASSENGER SERVICE had long been both the glory and curse of the New York, New Haven & Hartford. Its "Shore Line" main along the Connecticut coast was and still is the primary rail link between Boston and New York City, and its inland routes linked major New England cities to the network. Rapid acceleration was the secret to maintaining schedules on the curving Shore Line, and the New Haven was interested in high horsepower units.

In 1947 the New Haven hosted the Erie-Built demonstrators but didn't bite — the railroad was satisfied with its Alco DL109 fleet — but it was impressed with the CPA24-5 demonstrators of 1950 and not only bought eight new units (792-799) but bought the demonstrators themselves (790 and 791). On March 17, 1957, the 792 in factory green livery (opposite bottom) was paired up with an Alco PA1 at Stony Creek, Conn., and on August 2, 1959, another C-Liner/PA1 set (below) was westbound at East Haven with No.171, *The Patriot.*

In 1955 the New Haven tried three experimental light-weight trainsets for use on the Shore Line. One of the sets was comprised of Spanish Talgo cars (built in the U.S. by ACF) and two specially designed FM locomotives, one for each end of the train. The P12-42 units had eight-cylinder OP 1750-h.p. engines to provide head end power for the passenger cars and 1200 h.p. for traction, with two motors on the front trucks only. In the electrified territory of Grand Central Terminal they shut down the diesels and operated off the 600-volt d.c. third rail. The *John Quincy Adams* set was at Wickford Junction (opposite top) in September 1957. The Boston & Maine bought two more P12-42 units and a Talgo set, dubbed the *Speed Merchant,* for service between Boston and Portland, Maine. The rough-riding lightweights did not appeal to the public on either railroad and were soon withdrawn from high-speed service. B&M No.2 (opposite center) was at the scrappers in Leeds, Maine, on November 11, 1978.

A. HOLTL

NEW YORK, NEW HAVEN & HARTFORD

MODEL	NUMBER	B/N	DATE	NOTES
CPA24-5	790, 791	24L273, 24L274	?/50	Ex-FM demonstrators 4801, 4802; acq. 4/51
H16-44	560-564	16L279-16L283	11/50	Renumbered 590-594
H16-44	565	16L308	11/50	Renumbered 595
H16-44	566-569	16L315-16L318	11-12/50	Renumbered 596-599
CPA24-5	792-799	24L533-24L540	1/52	
H16-44	1600-1602	16L1032-16L1034	6/56	PC 5160-5162
H16-44	1603-1608	16L1042-16L1047	6/56	PC5163-5168
P12-42	3100, 3101	17L1058, 17L1059	1/57	Dual mode; equipped for 600-volt 3rd rail
H16-44	1609-1614	16L1126-16L1131	6, 8/56	PC 5169-5174

BOSTON & MAINE

P12-42	1, 2	17L1060, 17L1061	1/58	No dual mode equipment

THOMAS J. McNAMARA

54

Two flavors of H16-44s

THE H16-44 ROAD SWITCHER was actually two distinctly different locomotives, and the New Haven had clear examples of each. In April 1950 FM boosted the rating of its H15-44 by 100 h.p. to produce the first H16-44 (MKT 1591), but the basic locomotive carbody and Westinghouse electrical system were essentially unchanged. In November 1950 the New Haven began taking delivery of ten H16-44s in the classic Loewy carbody, set up long hood forward with FM-only m.u. capability and equipped with steam generators for passenger service. Renumbered 590-599, they were soon relegated to local freight and switching duties. The 591 was photographed (below) in Boston on February 20, 1960.

When EMD could not deliver GP9s quickly enough, the New Haven returned to FM for 15 more H16-44s in mid-1956. The 1600-1614 had the new Train Master style carbody, short hood forward operation and universal m.u. Equipped with steam generators, they were put into passenger service, such as the 1600 and a mate (left) at Berlin, Conn., on August 30, 1957. All eventually wound up in freight service, such as the pair (opposite bottom) sharing the Cedar Hill engine terminal in New Haven with ex-Virginian freight motors in June 1968. Only the 1600s were kept in service by Penn Central, and in August 1969 (above), the 5170 and 5161 were on a transfer to the Soo Line at Schiller Park, Illinois. ⊕

WILLIAM VOLKMER

 # Cabin Cruisers and the *Cannon Ball*

AMERICA'S BUSIEST commuter railroad sampled its first Fairbanks-Morse unit in June of 1949 when it borrowed PRR Erie-Built 9473. Impressed with the acceleration of the 2000-h.p. Erie, a year later the LIRR purchased six 2000-h.p. CPA20-5 cab units — the only order FM received for that model. The 2007 (above) was at Hicksville in August 1961. In 1950 the LIRR also tried FM H15-44 demonstrator 1503 and purchased it shortly thereafter; it was photographed (right) on a freight at Mineola on April 13, 1963. Desiring even more horsepower to replace the lusty former-PRR K4 4-6-2s on the heaviest trains, the LIRR bought four CPA24-5 cab units like the 2402, shown (opposite bottom) at New Hyde Park in September 1963. The crews liked the smooth ride of the B-A1A units so much that they dubbed them "cabin cruisers." In October 1951 the LIRR bought eight H16-44s in the Loewy carbody, bracketing their numbers around the lone H15-44 1503. The 1504 (opposite top) was at Morris Park engine terminal in December 1961, and the 1507 and 1508 (below) had the parlor-equipped *Cannon Ball* at Hicksville headed for Montauk in August 1961. ⊕

LONG ISLAND RAIL ROAD

MODEL	NUMBER	B/N	DATE	NOTES
H15-44	1503	15L14	6/50	Ex-FM 1500 (II)
CPA20-5	2001-2008	21L331-21L338	6-8/50	Only CPA20-5s built
H16-44	1501	16L373	10/51	
H16-44	1502, 1504-1509	16L407-16L413	10/51	
CPA24-5	2401-2404	24L497-24L500	3-6/51	

The Royal Blues

ALTHOUGH IT AVOIDED any Fairbanks-Morse road units, the Baltimore & Ohio acquired a modest fleet of switchers and road switchers. In late 1948 it jumped in for ten H10-44 switchers and followed up a little over two years later with ten H12-44s — these first 20 units were all in the Loewy carbody, as clearly shown on the 9704 and 9718 set (above) working Baltimore on October 22, 1967. Three years later, in 1970, the 9704 was one of eight H10-44s that the B&O sent to the power-starved Jersey Central on a lease/purchase agreement.

The 9720 (opposite top), shown working the lead south of the Camden Station train-shed in Baltimore in September 1968, was the first of the pair of spartan carbody switchers delivered in September 1953; five more similar units followed in April 1957.

In December 1952 the B&O bought two H16-44s, 906 and 907, that arrived in the Loewy carbody; renumbered 6700 and 6701, these two units were leased to the Jersey Central in 1967 as numbers 18 and 19 but returned to the B&O for retirement shortly thereafter. The rest of the B&O H16-44 fleet, eight units delivered between 1955 and 1957, were in the Train Master style carbody, as shown on the set idling in the twilight in Cincinnati (top) in the summer of 1968, led by the 6705 and 6702. By July 1970 the H16-44s had been renumbered again, and the 9742 and 9741 (right) were working Bayview Yard in Baltimore.

58

BALTIMORE & OHIO

MODEL	NUMBER	B/N	DATE	RENUMBERED	1969 RENUMBERING
H10-44	300-309	10L86-10L95	11-12/48	9700-9709	
H12-44	310-319	12L392-12L401	2/51	9710-9719	
H16-44	906, 907	16L697, 16L698	12/52	6700, 6701 (CNJ 18, 19)	
H12-44	196, 197	12L774, 12L775	9/53	9720, 9721	
H16-44	925, 926	16L961, 16L962	4/55	6702, 6703	9737-9738
H16-44	927	16L969	5-6/54	6704	6739
H12-44	9722-9726	12L1089-12L1093	4/57	Same	
H16-44	6705-6709	16L1144-16L1148	3/57	Same	9740-9743 (6709 not re#)

Communipaw

THE RAILROAD THAT PIONEERED dieselization in the United States when it purchased a GE/Ingersol-Rand 1000-h.p. boxcab switcher in 1929 had a long and productive relationship with Fairbanks-Morse following the August 1947 convention of the Association of American Railroads in Atlantic City, N.J. Among the units displayed there by FM was H15-44 demonstrator 1500, which caught the attention of the Jersey Central, which was well into dieselization with switchers as well as EMD and Baldwin cab units. Seeking a road switcher that could handle a variety of tasks, the CNJ found the H15-44 to be just what it was looking for and bought that demonstrator while placing an order for 13 similar units. When it arrived in September 1948 after overhaul at Beloit, the 1500 was CNJ's first road switcher, with Baldwins, Alcos

and EMD's to follow. The CNJ went on to purchase three H16-44s and a 13-unit fleet of Train Masters.

Communipaw engine terminal in Jersey City was a wondrous relic of the steam era, with twin turntables and round-houses — one for passenger engines and one for freight — and a huge concrete coaling tower. In the 1960s it was a gathering place for the diesels that had vanquished steam. In September 1966, the pioneer 1500 (right) was on a ready track alongside a Reading GP35 and GP30 which had arrived in pool service from Pennsylvania. Two years later (below) the scene was dominated by Train Masters, with an H16-44 in the right background. In April 1962 (opposite center and bottom) H16-44 1516 and Train Master 2409 showed off their "toothpaste stripe" livery in perfect poses at Communipaw.

Model	Number	B/N	Date	Notes
H15-44	1500	15L6	3/48	Ex-FM demonstrator 1500
H10-44	9700-9704, 9706	10L86-10L90, 10L92	11-12/48	Ex-B&O, Acq. 1970
H10-44	9708, 9709	10L94, 10L95	12/48	Ex-B&O, Acq. 1970
H15-44	1501-1513	15L15-15L27	1-4/49	
H16-44	1514-1517	16L304-16L307	7/50	
H24-66	2401-2407	24L849-24L855	5-6/54	
H24-66	2408-2413	24L885-24L890	3-4/56	

CENTRAL RAILROAD OF NEW JERSEY

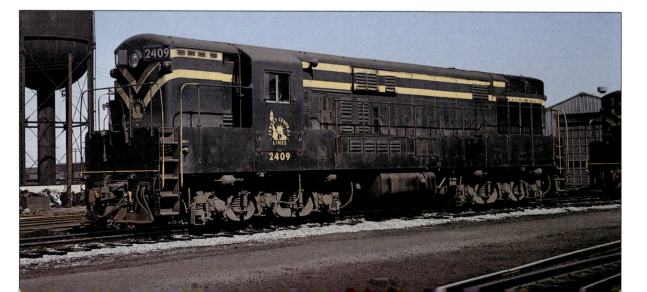

Jersey City Terminal

COMMUTER SERVICE demands power and acceleration, and CNJ was quick to recognize the potential of the 2400-h.p. Train Master. The railroad had H15-44s and H16-44s with boilers (except the 1500) and dual controls for use in commuter service, and the first seven similarly equipped Train Masters of 1954 forever sealed the fate of CNJ steam. Though designated long hood forward, the dual-controlled Train Masters were preferred by the crews to run short hood forward. In September 1966

(right) 2404 was leading an evening commuter train west out of Jersey City Terminal beneath its awesome array of semaphores. The next day, H15-44 1509 (below) was eastbound at Cranford, where the "Aldene" ramp was being built to route Raritan line trains onto the Lehigh Valley, beneath which the 1511 (opposite bottom) was westbound in December 1964. By June 1968 (bottom) trains were running out of Newark Penn Station instead of Jersey City; 2409 was westbound on the PRR at South Newark.

Old and New — Borrowed and Blue

BY THE MID-1960S, many of the Jersey Central Train Masters were being assigned to freight duty instead of commuter service. In August 1966, Train Master 2407 was teamed up behind one-year-old SD35 2504 (above) on an eastbound freight out of Allentown crossing the Delaware River between Easton, Pa., and Phillipsburg, N.J. In April that same year, H15-44 1510 (opposite top) was still adorned in its "toothpaste stripe" livery as it worked a local freight at the Manville/Finderne, N.J., station. The SD35s of 1965 were the last CNJ units delivered in the green and

yellow. Baltimore & Ohio management was in charge of the CNJ in December 1968 when 13 passenger GP40P's in B&O blue and yellow arrived to bump the last Train Masters off commuter service — in January 1969 the 2402 and 2401 (below) had a freight westbound along the Lehigh River at Jim Thorpe, Pa., with the Lehigh Valley main line visible on the opposite side of the river. In December 1967 the B&O began to lease power to the CNJ, beginning with nine nearly new SD40s. In 1970 the B&O sent up eight m.u.-equipped (cab end only) H10-44s, which were put to good use across the system. In April 1972 the 9703 and 9702 (opposite bottom) were idling outside the Bethlehem roundhouse.

READING COMPANY

Model	Number	B/N	Date	Notes	1967 Renumbering
Centercab	35 (re# 97)	1703	3/17/39	St. Louis Car Co. body with two 4-cycle 8-cylinder 300-h.p. engines	
H24-66	800, 801	24L779, 24L780	9/53	Freight: dynamic brakes, hump control , long hood front	801/201
H24-66	860, 861	24L781, 24L782	10/53	Passenger: boilers, dual controls, no dynamics, short hood front	None
H24-66	802-806	24L795-24L799	11/53	Freight: dynamic brakes and hump control , long hood front	806/202
H24-66	866, 867	24L863, 24L864	11/55	Passenger: boilers, dual controls, dynamics, short hood front	867/265
H24-66	862-865	24L865, 24L882-24L884	11-12/55	Passenger: boilers, dual controls, dynamics, short hood front	862-865/261-264
H24-66	807, 808	24L906, 24L907	12/56	Freight: dynamic brakes and hump control , long hood front	808/203

Rutherford and the *Queen*

AFTER A DISASTROUS introduction to FM in the form of No.97, a trouble-plagued 600-h.p. centercab powered by two four-cycle, eight-cylinder engines (page 5), it took the Reading a long time to return to Beloit. But the Reading was intrigued by the 2400-h.p. Train Masters of 1953 and ordered four of them (with the little 97 credited at $33,573.16 as trade-in) even before hosting the TM-1 and TM-2 demonstrators following the Railway Supply Manufacturers Association exhibition in Atlantic City in June 1953. By the end of 1956 the Reading had acquired a fleet of nine freight and eight passenger Train Masters. On August 6, 1962, the 863 had the *Queen of the Valley* (opposite bottom) departing Harrisburg; east of Allentown it would run over the CNJ to Jersey City. The CNJ often used its own Train Masters on the run-through passenger service pool. One month later (bottom) the 805 and 861 met at Belt Line Junction on the north side of Reading. In August 1966 (left) 860 was working the west lead at Rutherford Yard in Harrisburg, and in December 1968 the renumbered 265 (ex-867) was on Rutherford's east hump (below). Only 867 and 801 got the new green and yellow colors. ▲

KERMIT GEARY, SR.

Lackawanna Loewys

W.J. BRENNAN   MATT HERSON

THE DELAWARE, LACKAWANNA & WESTERN was "The Road of Anthracite," linking the hard coal fields of the Scranton, Pennsylvania, area with New York City to the east and Buffalo to the west. The Lackawanna began dieselizing rather rapidly in the late 1940s with a large fleet of EMD F-units for both freight and passenger service, and beginning in 1950 it turned to road switchers in the form of Alco RS3s, EMD GP7s and six FM H16-44s. The 930-935 packed Westinghouse electrical gear inside the Loewy carbody and were set up for long-hood-forward operation but were equipped with dual controls. Following the merger with the Erie on October 17, 1960, the units had 1000 added to their numbers, and many shed their simple black DL&W livery (shown **above right** on the 1931 at Denville, N.J., on the Dover Drill in August 1963) for the Erie-inspired Erie Lackawanna black and tan worn by the 1932 at Scranton (**above left**) in June 1965. Although they could m.u. only with each other, the H16-44s were often found in road sets, such as the 1930, 1933 and 1934 (**below**) at the power plant along the Delaware River in Portland, Pa., in January 1965. The 1934 was the only H16-44 to receive the newer EL maroon and gray livery. A four-unit set was westbound at East Stroudsburg, Pa., (**bottom right**) in September 1964. The Loewys were sold for scrap to Striegel in Baltimore in mid-1966 but were resold to the Chihuahua Pacifico in Mexico, where they worked for at least another half-dozen years (page 118) as CH-P 526-531.

KERMIT GEARY, SR.

DELAWARE, LACKAWANNA & WESTERN

MODEL	NUMBERS	B/N	DATE	EL NUMBERS	NOTES
H16-44	930-935	16L687-16L692	12/52	1930-1935	930-935 to CH-P 526-531 in 1966
H24-66	850-859	24L734-24L43	6/53	1850-1859	852, 859 to CH-P 534, 535 9/69
H24-66	860, 861	24L1035, 24L1036	11/56	1860, 1861	

"The *Erie* never had Train Masters!"

WHENEVER SOMEONE CASUALLY mentions the "Erie Train Masters," DL&W fan Mike Del Vecchio is quick to point out that, "the *Erie* never had Train Masters!" In fact, the Erie never had any FM's, and all the Erie Lackawanna FM's were of DL&W heritage. Impressed by FM's sales pitch for the new Train Master, in June 1953 DL&W placed an order for the first production H24-66s, even before demonstrators TM-1 and TM-2 rolled out of Beloit. The DL&W got ten brutes equipped with steam generators and wearing the maroon and gray passenger livery, as shown on the 853 (**opposite top**) at Paterson,

N.J., on July 3, 1957. Two more, 860 and 861, delivered in November 1956, were the next-to-last Train Masters built. Following the 1960 EL merger the Train Masters were renumbered and moved to the Cleveland-Brier Hill (Youngstown) iron ore service, where many received the Erie-inspired black and buff before the EL turned to the DL&W maroon for all of its power. In April 1964 EL 1852 (**below**) was idling the night away at Cleveland, while on June 17, 1968, the matched set (**above**) of 1853, 1855 and 1857 was working an ore train there. Four days later (**right**) 1852, 1861 and 1850 were doing the same thing. ⊕

ROBERT KRONE

The Virginian: Nearly 100% FM

VIRGINIAN RAILWAY

MODEL	NUMBERS	B/N	DATE	N&W NUMBERS
H24-66	50-57	24L807-24L814	3/54	150-157
H24-66	58-68	24L838-24L848	4-5/54	158-168
H16-44	10-15	16L866-16L871	6/54	110-115
H16-44	30	16L920	1/55	130
H16-44	17-27	16L921-16L931	12/54-2/55	117-127
H16-44	16, 31	16L932, 16L933	3, 2/55	116, 131
H16-44	32	16L948	2/55	132
H16-44	33	16L960	2/55	133
H16-44	28, 29	16L963, 16L964	12/54	128, 129
H16-44	34	16L975	10/55	134
H16-44	35-39	16L988-16L992	10/55	135-139
H24-66	69-73	24L1037-24L1041	5-6/57	169-173
H24-66	74	24L1048	6/57	174 (Last Train Master built)
H16-44	40-47	16L1132-16L1139	11-12/56	140-147
H16-44	48, 49	16L1177, 16L1178	10/57	148, 149 (Wreck rebuilds of 23 and 28; Last FM's for a U.S. customer)

FAIRBANKS, MORSE & COMPANY never had a more dedicated customer than the Virginian Railway. With the solitary exception of one GE 44-ton centercab switcher, every diesel on the Virginian was an FM. The Virginian had electrified its toughest territory in 1925 and bought modern steam power for the rest of the system, so when it was ready to dieselize in 1954 it could standardize to a degree that would be the envy of any other railroad. At that time only FM offered the 2400-h.p. six-motor that would give the Virginian performance equivalent to its electrics, and over the next four years the Virginian acquired a fleet of 25 of them, backed up by a 40-unit fleet of versatile H16-44s. The merger with the Norfolk & Western on December 1, 1959, ended both the electrification and the FM reign on the Virginian. But for one brief shining moment, the Mullens Motor Barn was Beloit's Camelot.

The N&W merger was almost a year away when H16-44s 32 and 46 (opposite bottom right) were photographed on December 12, 1958, in Roanoke. The N&W added 100 to the Virginian numbers, as shown on Train Masters 170 and 167 (top and opposite center) getting an N&W extra west rolling at VN Tower in Salem, Virginia, on April 1, 1960. Train Master 153 (opposite bottom left) was at Whitethorne, Va., on July 22, 1960. H16-44s 140, 144 and 148 (above) were still in their original colors as they worked on the N&W near Roanoke in August 1966 — note the N&W herald on the bridge wing wall. It is ironic that the Norfolk & Western, which never bought a new FM, would wind up with one of the nation's biggest fleets through mergers (pages 48-31, 74-79 and 104-109).

Banner Blues and Blacks

THE WABASH RAILROAD was FM's seventh customer, picking up the 21st and 22nd 1000-h.p. switchers produced at Beloit and following up with another pair two years later. That 22nd unit, Wabash 381 (**opposite bottom left**), was switching No.1, the Saint Louis-bound *Wabash Cannon Ball* — with an Alco PA and EMD E8 — at Decatur, Ill., on April 1, 1964. The last FM switcher purchased by the Wabash, H12-44 386 (**opposite center**), was photographed working the Decatur yard hump. In February 1954 the Wabash purchased the Train Master demonstrators TM-1 and TM-2 and two years later opted for six more. The 556 (**below**) was at Detroit's Oakwood Yard on January 27, 1962. In 1964 all of the Wabash Train Masters were re-engined by Alco with 16-cylinder 251B engines rated at 2250 h.p. and renumbered into the 590s as they were completed. Fresh out of Alco in new solid blue paint, the 597 (**right**) was ready to depart Decatur on May19, 1964, with two F7s and a fellow Alco Century 424. The re-engined demonstrator TM-2, now 599 (**opposite bottom right**) was at State Line Tower at Hammond, Indiana, in 1965. Note that the Wabash ordered low-level end platforms on its own Train Masters so that the crossover walkways would line up with other units, but the two former-demonstrators remained as-built.

H.G. GOERKE / MATT HERSON COLLECTION

WABASH RAILROAD

MODEL	NUMBER	1961 RE#	1964 RE#	B/N	DATE	N&W NO.	NOTES
H10-44	380, 381	—	—	10L21, 10L22	11/46	3380, 3381	
H10-44	382, 383	—	—	10L101, 10L102	3/49	3382, 3383	
H24-66	550, 551	—	598, 599	24L730, 731	4/53	3598, 3599	F-M Demonstrators TM-1, TM-2, Acq. 2/54
H12-44	384-386	—	—	12L744-12L746	3/53	3384-3386	
H24-66	552	—	592	24L891	4/56	3592	592-599 Re-engined by Alco in 1964
H24-66	552A	555	595	24L892	4/56	3595	
H24-66	553	—	593	24L893	5/56	3593	
H24-66	554A	556	596	24L894	5/56	3596	
H24-66	554	—	594	24L895	5/56	3594	
H24-66	556A	557	597	24L896	5/56	3597	

KERMIT GEARY, SR.

Nickel Plate + Wabash = N&W

THE NICKEL PLATE ROAD would take delivery of ten new 2-8-4s from Lima in the year following the December 1948 purchase of its first H10-44s. By that time the railroad already owned diesel switchers from Alco, EMD, Baldwin and Lima and eleven Alco PA1 passenger units but rostered not a single road freight unit. Bellevue, Ohio, was the focal point of the Nickel Plate, and the road's second FM switcher, 126 (opposite top), was there in June 1966. Berkshires were still handling fast freight when the last of the three orders of H12-44s was delivered in April 1958. That last unit, 155, was at Bellevue (top) with 148 in March 1967; all the NKP H12-44s had the spartan carbodies. Following the Norfolk & Western merger of October 16, 1964, the NKP units got 2000 added to their numbers, and many received the "Pevler blue" livery like the 2144 (above) in April 1971. Bellevue then became a haven for N&W's merger-partner FM's, like AC&Y H20-44 501 (opposite bottom right) in October 1968 and the Wabash Alco-repowered Train Masters. The 3593, 3594 and an unidentified sister (opposite center) were at the roundhouse in March 1967, and the 3593 and 3596 were working the hump (opposite left) in October 1968.

NICKEL PLATE ROAD

Model	Numbers	B/N	Date	N&W Nos.
H10-44	125-133	10L103-10L111	12/48-4/49	2125-2133
H12-44	134-138	12L721-12L725	3/53	2134-2138
H12-44	139-145	12L1082-12L1088	3/57	2139-2145
H12-44	146-155	12L1101-1110	3-4/58	2146-2155

HAROLD BUCKLEY, JR. / MATT HERSON COLLECTION

 # Pittsburgh & West Virginia — revisited

JAMES P. SHUMAN

PITTSBURGH & WEST VIRGINIA			
P&WV AND N&W			
MODEL	NUMBERS	B/N	DATE
H20-44	50, 51	L1038, L1039	10/47
H20-44	52, 53	20L33, 20L34	10/48
H20-44	54-59	21L51-21L56	1-2/51
H20-44	60-65	21L630-21L635	5/52
H20-44	66-71	21L709-21L712	2/53
H16-44	90-93	16L1140-16L1143	12/56-1/57

THE PITTSBURGH & WEST VIRGINIA was about as obscure as a heavy-duty railroad could get, riding the ridges and jumping the valleys between the Western Maryland at Connellsville, Pa., and the Wheeling & Lake Erie 111.5 miles northwest at Pittsburgh Junction, Ohio. But this important Pittsburgh bypass boasted huge 2-6-6-4 steam locomotives and a massively built right-of-way between the two coal-haulers with which it connected. After sampling a lone Baldwin VO-1000 switcher in 1943, the P&WV bumped the articulateds between 1947 and

1953 with 22 H20-44s, painted in a dazzling black, cream and red livery. They tried one Baldwin AS616 in 1951 and finished dieselization with four H16-44s in 1957. Back-to-back H20-44s 50 and 65 (left) were at an unidentified location on May 31, 1952, while H16-44 90 posed with the 50 (opposite top) at Rook Yard, south of Pittsburgh, on April 27, 1961. The P&WV was absorbed in the N&W-NKP merger of 1964, and its units kept their original numbers when the N&W transferred them to the Norfolk, Va., terminal area. H20-44s 66 and 59 (below), dubbed "P-Willies" by the tidewater crews, were at Norfolk on October 21, 1967.

While the P&WV is long gone, its colorful image lives on, having been adopted by the North Carolina Ports Railway for its five H12-44s, including ex-Army 1860 (above, at Morehead City, N.C., in March 1996) and three ex-Yankeetown Docks units (page 112). ⊕

C-Liners in the Rockies

CANADIAN LOCOMOTIVE COMPANY of Kingston, Ont., began building FM locomotives under license in 1951, and in May of that year it outshopped a pair of CPA16-4 passenger demonstrators, 7005 and 7006, named the *City of Kingston*. These units became Canadian Pacific 4064 and 4065. Eight more cab units and eight boosters for CPR were the total production of this Canadian-only model. Nearly identical CFA16-4 freight units soon followed. The 1970s found the FM's in freight service, concentrated in the Canadian Rockies, based out of Nelson, British Columbia, along with the numerous H16-44s. With Kootenay Lake in the background, in October 1972 H16-44s 8602, 8608 and 8711 were lined up (above left) at the Nelson enginehouse with CPA16-4 4057, while CFA16-4 4104 (below) posed there on May 20, 1974. CPA16-4 4052 and CPB16-4 4471 had GP9 8810 and CFA16-4 4105 on a freight (above right) at Shoreacres and skirting the West Arm of Kootenay Lake coming into Nelson (right) on July 13, 1970.

MATT HERSON

KEITH ARDINGER / MATT HERSON COLLECTION   MATT HERSON

MATT HERSON

CANADIAN PACIFIC RAILWAY

MODEL	NUMBER	CLC B/N	DATE	NOTES
CPA16-4	4064, 4065	2646, 2647	5/51	Ex-CLC demonstrators 7005, 7006, 4065 preserved at Ottawa
CPA16-4	4052-4057	2680-2685	6-9/52	4455 and 4456 preserved at Cranbrook as ex-Robot Cars
CPB16-4	4449-4454	2686-2691	6-9/52	
CFA16-4	4076-4081	2716-2721	4-5/53	
CFB16-4	4455-4458	2722-2725	4-5/53	
CPA16-4	4104, 4105	2726, 2727	4-5/53	4104 preserved at High River, Alberta
CPB16-4	4471, 4472	2728, 2729	4-5/53	
H16-44	8547-8556	2890-2899	6-7/55	8554 stored for preservation at Calgary, Alberta
H24-66	8900	2900	6/55	FM b/n 24L861
H24-66	8905-8910	2922-2927	6-8/56	8905 at Delson, Que., is only surviving Train Master
H24-66	8901-8904	2928-2931	8/56	
H24-66	8911-8920	2932-2941	9-10/56	
H16-44	8601-8610	2942-2951	11-12/56	
H16-44	8709-8728	2958-2977	3-6/57	8728 Last CLC unit built

Rockies II: Train Masters and H16-44s

TWO PHOTOS / MATT HERSON

THE FIRST TRAIN MASTER built for Canada was a Beloit demonstrator painted Canadian Pacific and numbered 8900. The railroad was impressed, buying the 8900 and ordering 20 more, which were built by the Canadian Locomotive Company of Kingston. The 8900 had a single steam generator for passenger service, but CPR ordered its first four (8901-8904) with dual steam generators for severe winter duty; their high short hoods were widened to the full width of the frame to accommodate the two large boilers. When the boilers were removed in 1960, conventional width high hoods were applied. The 8900 (left) was at Trail, B.C., on July 13, 1970, while 8912 (below left) was at Port Coquitlam, B.C., on March 2, 1963.

The Train Masters and H16-44s were built to run long hood forward, but CPR later swapped the control stands for short-end-forward orientation and reversed the gray-nose paint job. The 8552 (opposite center), working the smelter at Trail on July 13, 1970, wore the long-hood-front paint, while 8721 and 8723 (top), two days later at Harrogate, B.C., had the short-hood-front scheme. Action Red CP Rail 8720 (opposite top) was working out of Nelson, B.C., in June 1974.

CN C-Liners in London

IN SEPTEMBER 1965 the Canadian National was just about to shake up its passenger service with updated schedules, new equipment and a positive image policy — but for the moment, the traditional steam-era schedules were intact, even though the paint job was modern CN. And the motive power was quite remarkable. London was a junction in the middle of Ontario where the Toronto-Hamilton-Chicago main line via the Sarnia-Port Huron Tunnel crossed the Toronto-Stratford-Windsor/Detroit secondary main. In the early afternoon two Toronto-bound trains would cross paths and exchange passengers at London. On this day, No.106 out of Windsor rolled grandly into town (right) over a massive truss bridge behind CPA16-5 6703 and Montreal/Alco FPA4 6772. A few minutes later, No.6, the *Inter-City Limited* out of Chicago via Port Huron pulled in behind Grand Trunk Western Geeps (opposite center). After a ten-minute transfer of patrons, No.6 departed for Toronto via the Hamilton line and was followed about five minutes later by 106 with its FM and MLW to run the more northerly route via Stratford. The next day, 106 departed London (opposite bottom right) with MLW FPA2 6758 leading CPB16-5 6805; the 6758 was one of the two prototype "FPA4s" created in 1958 by replacing their Alco 244 engines with new 251B's.

The CPA16-5's impressive A1A rear truck (below) made a compelling photo study in the Windsor station. The B-A1A configuration was used to distribute the weight of the train-heating steam generator. CNR had purchased six cab-and-booster sets of these passenger C-Liners in late 1954, and by 1966 they

were still working the fast trains in the Toronto-Windsor pool in the company of the FPA4s. The CPA16-5s had been delivered in CNR's green, yellow and black livery, shown on the 6704 (below) at Spadina round-house in Toronto in April 1961. The 1600-h.p. passenger C-Liners were sold only in Canada (CPA16-5s to CNR and CPA16-4s to CPR), with U.S. customers opting for the 2000-h.p. (LIRR) or 2400-h.p. (LIRR, New Haven, NYC) versions. Both CNR and CPR also had 1600-h.p. freight C-Liners. 🚂

MATT HERSON

Canadian National

Model	CLC B/N	Date	Numbers	Renumber
CFA16-4	2648, 2649	1/52	8700, 8702	9300, 9302
H10-64	2650-2652	10/51	7615-7617	1615-1617
H10-64	2653-2667	10/51-1/52	7600-7614	1600-1614
CFA16-4	2668	5/52	8704	9304
CFB16-4	2669-2671	5/52	8701, 03, 05	9301, 03, 05
H12-64	2672-2679	1/53	7622-7629	1622-1629
H12-64	2692-2695	8-9/52	7618-7621	1618-1621
CFA16-4	2696-2715	12/52-3/53	8706-8744 (even)	9306-9344 (even)
CPA16-5	2850-2855	12/54-2/55	6700-6705	
CPB16-5	2856-2861	12/54-2/55	6800-6805	
H16-44	2862-2879	3-5/55	1841-1858	2200-2217
H12-44	2880-2889	8-10/55	1630-1639	
H24-66	2901 (FM 24L862)	7/55	3000	2900
H12-44	2902-2921	4-6/56	1640-1659	

Unique hoods and a lone Train Master

THE FIRST FM UNITS built for the Canadian National were eighteen unique 1000-h.p. light road switchers in 1951 riding on A1A trucks (and even the trucks were unusual, having the center idler wheels of smaller diameter than the motor-driven outer wheels). They were designed for branch lines with light rail where loadings could not exceed 40,000 lbs. per axle. Twelve more, rated at 1200-h.p., followed a year later, and the H10-64s were later upgraded to 1200 h.p. H12-64 1626 (top) was at Moncton, N.B., in July 1966. At the time the H10-64s were being delivered, CNR acquired the first of 23 CFA16-4 freight C-Liners; 9300 (opposite top) was at Quebec City on February 8, 1963. CNR's 20 H12-44s differed from the U.S. models in having road trucks, ten-inch longer carbodies and raised side walkways; the 1659 (right) was photographed in Montreal. The only "conventional" FM's built by Kingston for CNR were the eighteen H16-44s like the set led by 2217 (right center) at Cantic, Quebec, in October 1962. The Kingston builders plate 2875 (opposite center) was on H16-44 2213 at Sarnia, Ont., in October 1965.

In mid-1955 Beloit turned out two Canadian demonstrator Train Masters, one for CPR and one for CNR. The CNR bought "its" 3000 but never came back for more. Renumbered 2900, the lone H24-66 worked passenger service (left, at Richland, Ont., in February 1963), until reassigned to hump duty. "CNR never knew quite what to do with it," observed Canadian fan Ken Goslett. ⊕

A Tale of Two 100s

THE SANDERSVILLE RAILROAD is one money-making machine. Beneath the mid-Georgia red clay are deposits of kaolin, a chalk-like mineral which is used to make paper white, and a significant percentage of the kaolin used in the U.S. and Canada originates on the Sandersville's nine miles of railroad. In August 1953 the Sandersville retired its steam power with a brand new H12-44, number 100. In 1964 they went to EMD for an SW1200, number 200, and in late 1967 they were ready for another EMD, an SW1500. I was a new field instructor working for EMD and was assigned to the delivery of the Sandersville unit. Since the Tarbutton brothers, whose family owned the railroad, had apparently decided to go for an all-EMD locomotive fleet, they specified the number of the new unit to be 100. When the SW1500 arrived in Sandersville on Monday, January 15, 1968, the railroad had three locomotives on the property: numbers 100, 200 and 100 (top). The next morning, FM 100 was spotted in front of the yet-to-be-put-into-service EMD 100 (opposite top) while the 200 went out to earn some money by hauling kaolin to the CofG interchange. With the help of a laborer, General Manager L.E. Griffith (above) began to resolve the number conflict by simply painting out the last zero from the FM's number, creating Sandersville No.10. Later that week the whole fleet ganged up on the Second Switch Job (opposite bottom): SW1500 No.100, SW1200 No.200 and H12-44 No.10. That trio would serve the railroad until 1970, when another new SW1500, No.300, would retire the FM.

SANDERSVILLE RAILROAD

MODEL	B/N	DATE	NUMBER
H12-44	12L777	8/53	100 (re# 10 1/15/68)

TOM SMART / MATT HERSON COLLECTION

LOOK AHEAD·LOOK SOUTH

THE SOUTHERN RAILWAY got its first OP engines in six railcars in 1939 (page 4) and followed up in 1950 and 1951 with a fleet of sixteen Loewy-carbodied H16-44s, all delivered in the classic steam-era green and gold livery. By the early 1960s most had traded their green for the black "tuxedo" image, worn by the 2151 and 2147 (left) at Louisville, Ky., in August 1962. The 6548 (bottom right), at East St. Louis a year later, was numbered and sublettered for subsidiary Alabama Great Southern.

In mid-1955 the Southern bought five Train Masters for the Chicago, New Orleans & Texas Pacific "Rat Hole" line from Cincinnati to Chattanooga. The monsters had Wabash-style low end platforms and walkovers, as shown on the green 6301 (top right) and were most frequently seen sandwiched between a pair of F-units, such as 6304 (below) at Sunbright, Tenn., in October 1955.

SOUTHERN RAILWAY

Model	Number	B/N	Date	Notes
Motor Car	SR 1-4; AGS 40, 41	StLCCo. 1598	1/39	F-M 800 h.p.O.P. engines
H16-44	SR 2146-2150	16L310-16L314	12/51	
H16-44	AGS 6545-6550	16L319-16L324	12/50	
H16-44	SR 2151-2155	16L511-16L515	12/51	
H24-66	CNO&TP 6300-6304	24L856-24L860	5-6/55	

R.D. SHARPLESS / WALTER A. APPEL COLLECTION

Georgia, Texas & Arizona

THERE ARE MANY RAILROADS that would not immediately come to mind at the mention of Fairbanks-Morse. Take, for instance, the Central of Georgia, which rostered five Loewy H15-44s that traded in their fancy gray, blue, black and orange factory paint for a more sedate Pullman green and yellow. The units were limited by their ability to multiple only with each other and only at the short (rear) hood end, as shown (**below**) on the 105-104 set at Chattanooga in December 1965. The CofG also had four H12-44s like the 316, which was photographed (**right**) on the Monon at Salem, Indiana, on February 5, 1966, en route to EMD in La Grange as a trade-in on new power.

TOM SMART / MATT HERSON COLLECTION

J. DAVID INGLES / MATT HERSON COLLECTION

ANOTHER OBSCURE OWNER of just five FM road switchers was the Missouri-Kansas-Texas, which had them in two separate classes! In June 1949 the Katy got one freight H16-44, 1591, and followed up a little over a year later with four boiler-equipped H16-44s for local passenger service in Texas and Oklahoma. By March 1964, passenger 1732 had lost its boiler and was photographed in Fort Worth (**opposite bottom**) as the renumbered 159.

THE APACHE RAILWAY in Arizona replaced its steam locomotives between 1947 and 1952 with two FM H10-44s, Nos. 100 and 200, and one Baldwin S12 switcher for the 72-mile

CENTRAL OF GEORGIA			
MODEL	NUMBER	B/N	DATE
H15-44	101-105	15L30-15L34	6/49
H12-44	315-318	12L726-12L729	3/53

run between the Southwest Forest Industries' mill at McNary and the Santa Fe connection at Holbrook. As business picked up, the Apache became famous for its fleet of Alco road switchers, but the two FM's remained in service until 1978. On February 18, 1975, the 200 (**opposite top**) was in its traditional colors in the snow at Snowflake, while on March 19, 1976, the appropriately numbered unit (**opposite center**) was wearing a new Bicentennial livery.

92

KEITH E. ARDINGER / MATT HERSON COLLECTION

APACHE RAILWAY

MODEL	NUMBER	B/N	DATE
H10-44	100	L1081	10/47
H10-44	200	10L100	11/48

MISSOURI - KANSAS - TEXAS

MODEL	NUMBER	RE#	B/N	DATE
H16-44	1591	157	16L369	4/50
H16-44	1731-1734	158-161	16L516-16L519	9-10/51

JIM MARCUS / MATT HERSON COLLECTION

DENNIS E. CONNIFF III / RICHARD R. WALLIN COLLECTION

Black & Gold and Red all over

BACK IN 1948 there were some strong similarities between the H10-44s of the Denver & Rio Grande Western and its Midwestern neighbor the St. Louis-San Francisco. Both the Rio Grande and Frisco switchers had the classic Loewy carbodies and all were delivered in black with yellow "scare stripes" — though the D&RGW referred to its color as "Grande gold." Frisco 273 (below) was at West Tulsa, Okla., on November 28, 1964, while Rio Grande 123 (opposite center) was at Roper Yard in Salt Lake City in July of that same year. In the mid-1960s the Frisco adopted its red and white livery (inspired by EMD's GP35 demonstrators) as worn proudly by the 280 (right) in June 1973; note the Frisco herald beneath the FM medallion on the nose.

In January 1967 the Rio Grande traded H10-44s 120 and 121 to EMD on SD45s and sold the 122 and 123 a year later to Precision Engineering of Mt. Vernon, Ill., for scrap. Precision then resold the 122 to the Frisco, where it got the red colors as 286, shown (opposite bottom) in July 1970.

The Rio Grande's very first road switchers were three H15-44s, delivered in January and February 1948, just two months ahead of the H10-44s. They were used as helpers on Soldier Summit and in local freight work before settling into yard duty at Grand Junction, where the 152 was photographed (bottom) on October 10, 1965. The 151 and 152 were set up for use with "auxiliary booster" (hump slug) 25, a cut-down Baldwin VO660.

MATT HERSON

GORDON E. LLOYD / RICHARD R. WALLIN COLLECTION

MAC OWEN / MATT HERSON COLLECTION

P.D. CUSTER / ALAN MILLER COLLECTION

DENVER & RIO GRANDE WESTERN

MODEL	NUMBER	B/N	DATE	NOTES
H15-44	150-152	15L3-15L5	1-2/48	
H10-44	120-122	10L53-10L55	4/48	122 to SLSF 1969
H10-44	123	10L58	6/48	

ST. LOUIS - SAN FRANCISCO

H10-44	270-275	10L47-10L52	3/48	
H10-44	276-278	10L83-10L85	9-10/48	
H10-44	279	10L99	10/48	
H10-44	280, 281	10L132, 10L133	6, 7/49	
H12-44	282-285	12L433-12L436	6/51	
H10-44	286	10L55	4/48	Ex-D&RGW 122

KEITH E. ARDINGER / MATT HERSON COLLECTION

G.J. BOLINSKY / KEITH E. ARDINGER COLLECTION

Weyerhaeuser

THREE PHOTOS: KEITH E. ARDINGER

THE WEYERHAEUSER TIMBER COMPANY gathered an interesting collection of FM switchers for use on its numerous logging operations in the Pacific Northwest. In April 1948 it bought H10-44 No.481 for use on its Vail McDonald Branch, a 30-mile line haul southwest of Tacoma to a Puget Sound log dump. The adjoining Chehalis Western, another 30-mile road haul out of Pe Ell, Wash., got two H10-44s in 1949, the 492 and 493; the 492 (**opposite top**) was at the Vail shop on June 25, 1976. The last H10-44 was the "D-1" for the Columbia & Cowlitz, the railroad that works the huge lumber mill complex at Longview, Washington. In 1968 the D-1 was traded to Alco on Century 415 No.701, and in May 1969 was sold by Alco to the Pacific Great Eastern, where it was photographed (**right**) at North Vancouver, B.C., in November 1971. In April 1972 the PGE was renamed British Columbia Railway, and the D-1 was renumbered 1004 a year later.

In August 1951 Weyerhaeuser got its first H12-44, No.1, for the White River Lumber Division at Enumclaw, Wash., where it was photographed (**left center**) on August 4, 1969. The only spartan carbody switcher owned by Weyerhaeuser was Columbia & Cowlitz H12-44 D-2, which was renumbered 700 and photographed on the trestle at Longview (**top**) in June 1970. It had been repainted blue by the time it was photographed at Longview (**above left**) on April 9, 1982. White River Timber No.1 was sold to Pacific Transportation Services as 121 and was (**left**) at Tacoma on August 25, 1982. The only survivor of Weyerhaeuser's fleet, it resides today at the museum in Snoqualmie. ⓕⓜ

WEYERHAEUSER TIMBER COMPANY

MODEL	RAILROAD	NUMBERS	B/N	DATE	NOTES
H10-44	Vail McDonald Branch	481	10L60	4/48	Scrapped.
H10-44	Chehalis Western	492, 493	10L148, 10L149	5, 6/49	492 became PTS 122, scrapped; 493 scrapped
H10-44	Columbia & Cowlitz	D-1	10L174	6/49	Traded to Alco 1968 on C415 701; to BCR 1971, re# 1004 6/72, scrapped 11/75
H12-44	White River Timber	1	12L437	8/51	Re# 714, to Pacific Transportation Services 121, to Puget Sound Ry. Hist. Assn.
H12-44	Columbia & Cowlitz	D-2	12L1025	5/56	Re# 700, scrapped

Train Masters by the Bay

IT WAS THE GRANDEST show of Fairbanks-Morse power in North America. Every weekday afternoon scarlet and gray Train Masters would line up five or six abreast alongside the umbrella sheds of the Southern Pacific passenger station at Third and Townsend in San Francisco, ready for the rush hour "commute" parade down the peninsula to San Jose. In October 1971 (above and top) trains 130, 132, 136, 128 and 134 were lined up left to right — adding the big "SP" to the noses in 1970 obliterated the locomotive numbers which had previously occupied that space, making it impossible to pick out unit numbers from such front-only views, since the train numbers occupied the illuminated train indicator boards.

Following tests with the "Western" Train Master demonstrators TM-3 and TM-4 in mid-1953, in November that year the SP placed an order for 14 Train Masters and bought the two demonstrators. The short time between the order and the December delivery of the 4802-4809 lends credence to the report that those eight units were already under construction at Beloit as New York Central 4600-4607 when the order was cancelled. The boiler-equipped Train Masters were acquired by SP for use not in commute service but for freight and passenger duty on the Texas lines. The brand new 4809 and 4807 (right) were at El Paso on January 3, 1954. These were SP's first hood units to get the "Black Widow" road unit livery. ◉

SOUTHERN PACIFIC				
MODEL	B/N	DATE	NUMBERS	RENUMBER
H12-44	12L653, 12L658	9-10/52	1486-1491	2350-2355
H24-66	24L732, 24L733	12/53	4800, 4801	3020, 3021 (ex-demos TM-3, TM-4)
H12-44	12L760-12L766	5/53	1529-1535	2356-2362
H12-44	12L771-12L773	8/53	1536-1538	2363-2365
H24-66	24L787-24L790	2/54	4810-4813	3030-3033
H24-66	24L791-24L794	12/53	4802-4805	3022-3025
H24-66	24L800, 24L801	2/54	4814, 4815	3034, 3035
H24-66	24L803-24L806	12/53	4806-4809	3026-3029
H12-44	12L833-12L836	1/56	1568-1571	2366-2369
H12-44	12L976-12L980	1-2/56	1572-1576	2370-2374 (1575, 76 ordered as T&NO 119, 120)
H12-44	12L1062-12L1081	10/56-2/57	1577-1596	2375-2394

S·P The Peninsula commutes

AFTER ENCOUNTERING problems with their air filters in the Texas, Arizona and New Mexico desert country, the Train Masters were moved in the summer of 1956 to California's Coast Division for general freight and passenger service. Espee historian Joe Strapac noted that as the 4-8-4s, 4-8-2s and 4-6-2s in the commute pool came due for expensive maintenance, they were replaced by Train Masters, and steam was completely gone by late January 1957. The Train Master fleet had found its permanent home. Between 1960 and renumbering to the 3020-series in 1962, the Train Masters received the SP's new scarlet

and gray livery. In August 1970 the 3024 (above) had train 147 northbound a Millbrae with a typical train of old round-roof Harriman coaches and a new bi-level gallery car. Demonstrator TM-4 originally had no steam generator, but one was installed when it was sold to SP; it was working as the 3021 (below left) at South San Francisco in August 1970, the same month 3031 was photographed on train 120 (below) exiting one of the numerous tunnels on the south side of San Francisco. In July 1969 the 3033 was bringing train 138 (opposite top) into the San Jose station, 46.9 miles south of Third and Townsend. The 3022 (opposite bottom) was idling at the San Jose round-house that same day. The commute Train Masters were turned at each end of their runs and always ran short hood forward. It was twilight in November 1971 as train 134 (opposite center) rumbled through Burlingame; by now gallery cars were outnumbering the Harrimans. 🔄

California switchin'

THE FIRST H12-44 to be built with the new "spartan" carbody was Southern Pacific 1486 in September 1952, shown (below) at Bayshore roundhouse on May 17, 1960. The only holdover of the Loewy styling was the rounded fillet between the front of the cab and the raised walkway, also evident on the 2351 (opposite center), originally 1487, at South San Francisco in June 1974. Between 1952 and 1957 the SP received a total of 45 H12-44s in six separate deliveries. The 2387 (above), one of the last group, was at San Luis Obispo in July 1969. Although they were all put into service at El Paso when delivered, about two-thirds of the fleet migrated to the Coast Division, where they benefitted from the same experienced shop forces that kept the Train Masters running in commute service (the rest remained in service around Tucson and El Paso). The 2372 (opposite top) was on a local freight at South San Francisco in August 1970, while the 2381 (opposite bottom) was working the passenger station during an afternoon rush in October 1971, clattering through the lead switches at the 4th Street Tower. All SP FM's were retired in the mid-1970s, and the only survivor is H12-44 2379, which was sold to Southwest Portland Cement in 1974 and used at Odessa, Texas, before being donated to the Age of Steam Railroad Museum in Dallas.

ALAN MILLER COLLECTION

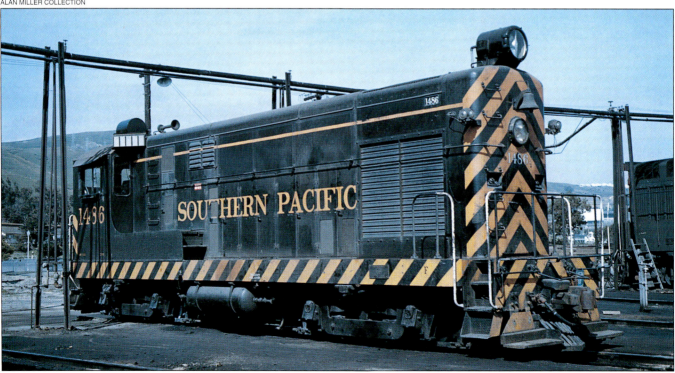

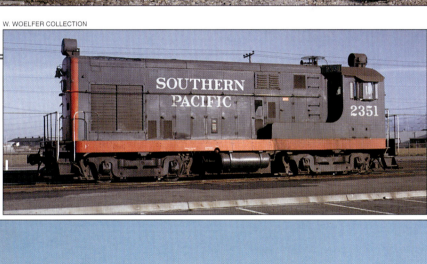

Elmore in the autumn

NORFOLK & WESTERN, which never bought a new FM unit for itself, ended up with one of the all-time largest fleets by merging with the Virginian, Nickel Plate, Wabash, P&WV and AC&Y for a total of nearly 140 units. And nowhere did the N&W put on a more impressive show of OP power than on the old Virginian, deep in the hollows around Mullens, West Virginia. Elmore Yard stretches along the Guyandotte River east from Mullens, and in October 1973 the "3598" (left) was working Elmore. This was actually former Virginian/N&W 153, the third of four units to wear N&W number 3598 in order to fulfill the financial obligations for the Alco re-engined Wabash unit, which had been disposed of. The suspension footbridge (opposite bottom) to the east end yard office spans the Guyandotte. In May 1975 the 158 (below and bottom) was working the east end of Elmore Yard, filling the hazy valley with the distinctive sound of its OP engine. 🖐

 # Oak Hill and Page

THE "OAK HILL TRAIN MASTER" was a railfan legend of the 1970s. The old Virginian White Oak Branch was a small railroad shaped like a "T" laid down on its right side. At the base was the VGN Mullens-Deepwater main line, at the crossbar intersection was the Oak Hill depot, at the south bar of the "T" was the C&O connection at Carlisle and at the north crossbar was Lochgelly No.2 Mine at Summerlee. The N&W kept Train Master 171 at

Oak Hill, specially equipped with a crankcase heater that could be plugged into an electrical outlet at the depot to permit the engine to be shut down overnight in the winter, as it was doing (opposite top) in January 1970. Earlier that day, sister 153 and two other Train Masters (below left) had to be kept idling in the snow a few miles north in the small coal collection yard at Page. In October 1973 the 171 was at Carlisle (below right) working the

stub-end yard alongside the New River Company store in the neighborhood known as Oakwood. At Carlisle the 171 picked up raw "cleaner coal" off the C&O to be taken to the preparation plant at Summerlee (opposite top), where it would be cleaned, screened and graded before being reloaded for shipment to a customer. In late October 1970, the 171 was working the yard in Oak Hill (two photos below right). In October 1972 (below left) the 171 was making a caboose hop on the line to the Oak Hill Junction interchange with the Deepwater main line. The 171 was replaced in 1975 with an ex-NKP SD9, the Summerlee prep plant closed a year later, and the Oak Hill job was abolished. In December 1975 the 171 was cut down by the Roanoke Shops and converted into hump slug 9913.

N&W FM Farewell

MOST RAILROADS RETIRED their FM locomotives in relative obscurity. It was different on the N&W because of the size and variety of its fleet and the lateness of the date. By the early 1970s FM's were recognized as being a rare and endangered species. Fans flocked in from all over the country to see the "Mullens motor barn" full of Train Masters, like the 156 (left) in August 1971. But the N&W had practical uses for the FM's in its fleet even as they were being retired. New Alco Century 630s 1135-1139 were built in September 1967 with trucks from retired Train Masters, as shown on the 1137 (below) at Elmore on November 3, 1973, and at Hagerstown, Md., (bottom left) in December 1967. An even more valuable use for retired Train Masters was created when the Roanoke Shop cut them down to make tractive effort "slugs" out of them so that one high horsepower unit like a Century 630 could power the traction motors on the engineless slug and efficiently replace two or three low-horsepower units in slow, heavy coal yard work at places like Roanoke and the tidewater port of Lamberts Point, Virginia. Slug 9906 (bottom right) at Lamberts Point in March 1980 was originally Virginian 67, rebuilt by Roanoke in November

KURT R. REISWEBER

1974. Previously employed in the same duty had been H16-44s like ex-Virginian 119 and 118 (**below right** in November 1972) and AC&Y 205 (**below left** in February 1970). Twenty Train Masters were converted into slugs by Roanoke, and as a result, not one N&W Train Master was saved for a museum.

The N&W did send its Train Masters out in style, however. On November 3, 1973, the two newest Train Masters, Virginian 73 and 74, outshopped by Beloit in June 1957, headed up an NRHS passenger excursion from Roanoke to Mullens. Power originally slated for the train was 172 and 173, and both were repainted in fresh black with the new "NW" herald. The 172 failed just before the trip, however, and the more historically significant 174 — still in its working livery — was substituted. Dropping down from Clarks Gap on the dreary day, the 174 led the train beneath the remaining catenary supports on the the steel trestle at Bud (**above**), recreating one of the most famous images of

the Virginian electrified territory. In a bit of irony, as the trip departed Mullens (**opposite center**) it met C630 1137 at Elmore, with the Alco riding on Train Master trucks. The 173 became the last N&W Train Master in service, working the second shift at Elmore Yard on June 30, 1976. It was teamed up with Bicentennial SD45 1776 for excursions out of Roanoke on the July 3-4 weekend and was sidelined immediately thereafter. That month the 174 was converted into slug 9914, and the 173 met the same fate in July 1981 as it became slug 9919.

Even though I was certainly no stranger to FM's, the November 1973 Mullens trip was the first time I really "listened" to an OP; I d never consciously analyzed their sound before and was expecting sort of a "smooth Alco" sound. I was rather flabbergasted to realize that it was much more like that of a non-turbocharged EMD — sort of an overgrown Geep! It figures: they're both two-cycle engines. It was a valuable lesson learned just in time.

JEREMY PLANT

OP's on the water

FRANK ETZEL

ALTHOUGH FAIRBANKS-MORSE did have locomotive use in mind as it was developing its opposed-piston engine, it was a specification bid sheet put out in 1932 by the U.S. Navy for submarine power plants that brought the design into sharp focus. As the spectre of World War II appeared on the horizon, the OP became the most widely used engine in the U.S. submarine fleet. A fine example of a wartime sub is the *U.S.S. Ling*, built by the Wm. Cramp & Sons Shipyard in Philadelphia and launched on August 15, 1943, as a diesel-electric boat packing four 9-cylinder OP's rated at 1535 h.p. each. It's on display today in the Hackensack River at Teaneck, N.J. The tide was low on December 1, 1994, (**right**) as New York, Susquehanna & Western Dash 8-40B's led a stack train across the river behind the *Ling*.

The OP's continue to be used in other marine applications like the Ohio River Company's towboat *Harlee Branch, Jr.*, shown (**opposite top** and **center**) assembling its tow at O/Kan Harbor at Point Pleasant, W.Va., on December 4, 1969. Built by Dravo in

Pittsburgh in 1966, it is powered by two 12-cylinder OP's.

For its coal port operations in Newport News, Va., the Chessie System employed four FM-powered tugboats, all 110-footers designed by Thomas Bowes for New York Harbor service and built at the RTC Shipyard in Camden, N.J., in the 1950s originally for the CNJ and B&O. Each is powered by a

10-cylinder OP (**right**, in the *J. Speed Gray*) rated at 1600 h.p. and driving a fixed-pitch propeller through a reverse-reduction gearbox. The *Howard E. Simpson*, *M.I. Dunn* and *J. Speed Gray* were working the export coal carrier *Havfalk* (**below**) on April 1, 1983. The Chessie tugs were replaced by two new EMD-powered boats in 1984 but were sold for use in the Northeast.

Yankeetown's white coal haulers

GEORGE HORNA / MATT HERSON COLLECTION

YANKEETOWN DOCK CORPORATION

MODEL	B/N	DATE	YDC NUMBER	PREVIOUSLY
H12-44	12L776	8/53	1	FM Demonstrator 76, to. YDC 5/54
H12-44	12L1024	5/56	2	
H12-44	12L1100	4/57	3	Ayrshire Collieries 1, Thunderbird Collieries 1

BACK WHEN YOU could find FM's in just about any American city, few railfans would bother with obscure places like southern Indiana coal mines. But by the early 1970s, FM's anywhere were recognized as being significant, and Boonville, Indiana, had more than its share. In 1953 Ayrshire Collieries and Peabody Coal had teamed up to build an eleven-mile railroad south from the Southern Railway at Boonville to link two new strip mines with a barge loader on the Ohio River at Yankeetown. In 1954 the Yankeetown Dock Corporation bought FM H12-44 demonstrator No.76 as the railroad's first locomotive, No.1. In mid-1956 H12-44 No.2 was added, and a year later Peabody extended the line ten miles further north to its mine at Lynnville, on a branch of the New York Central. Ayrshire's Thunderbird Collieries H12-44 No.1 later rounded out the fleet as YDC No.3. Yankeetown No.1 was at the Lynnville Mine (right) on September 17, 1973, and was teamed up behind No.2, working at Lynnville

112

(right and below) in August 1971, meeting the Penn Central job down the branch from Buckskin. In May 1977 No.2 was southbound just above Yankeetown (left). Thunderbird No.1 was fully repainted as YDC No.3 at Yankeetown (opposite top) on February 17, 1973; by then the FM's had been bumped from most road duty by two new SD38-2s. In 1996 the three were sold to the North Carolina Ports Railway (page 79) for service. 🚂

 # The "orphan" Babies

THE WHITE SWITCHERS of the Yankeetown Docks were not the only FM attractions of Boonville, Indiana. Alcoa Aluminum had built a large smelter adjacent to the barge facility at Yankeetown and used the Docks main line to reach its Squaw Creek Mine north of Boonville. To handle the coal between the mine and the power plant at the smelter, in January 1958, Alcoa purchased the only Baby Train Master built since the fleets for the C&NW and Milwaukee Road. With its full number 721001 shortened to "001" in the number boards, the Squaw Creek FM was southbound with loads (below) just above Yankeetown in July 1973 and was crossing the highway two miles south of Boonville (above) on April 8, 1974. In 1977 Alcoa put its "orphan" Baby into storage at the Lynnville mine and replaced it with two ex-Santa Fe Alco DL600B "Alligators," 9842 and 9843. In May 1977 the 9843 (opposite top) was southbound at Boonville. In 1983 the 721001 was sold to John Burbridge and is on display today at High River, Alberta, in the guise of "CPR demonstrator 7009." ⊕

PETER BERGS

JEREMY PLANT

THE LAST LOCOMOTIVE built by FM for a U.S. customer was another Baby Train Master: Tennessee Valley Authority 24, built in October 1958 and followed only by Mexican H16-44s. The 24 (renumbered F3060 in 1991) spent its entire life switching coal cars at the Gallatin Steam Plant in Tennessee, and until September 1995 it wore an uninspired black livery. At that time, it was decked out in the Gallatin plant's new blue and gray colors. It was hard at work pushing CSX hoppers (**above**) toward the rotary dumper on December 26, 1995.

Prior to the arrival of the 24, Gallatin had been worked by H12-44 No.22, which was transferred to the Shawnee Steam Plant near Paducah, Kentucky. The 22 was sold to the American Milling Company at Cahokia, Ill., and suffered an engine fire before this photo (**right**) was taken on September 3, 1994. Amazingly, it survives, having been bought in 1996 by the North Carolina Ports Railway.

Model	Number	B/N	Date	Notes
TENNESSEE VALLEY AUTHORITY				
H12-44	22	12L778	11/54	Shawnee, Ky., steam plant
H16-66	24 (F3060)	16L1157	10/58	Gallatin, Tenn, steam plant
ALUMINUM COMPANY OF AMERICA				
H16-66	001	16L1159	1/58	Squaw Creek Coal 721001

Ferrocarril de Chihuahua al Pacifico

Model	Number	B/N	Date	Notes
H16-44	501-504	16L943-16L946	7-8/55	
H16-44	505, 506	16L970-16L971	7/55	
H16-44	507	16L1002	2/56	
H16-44	508	16L1158	11/57	
H16-44	509-513	16L1181-16L1185	2-3/59	
H16-44	600-602	16L1186-16L1188	12/60	600-604 boiler equipped
H16-44	514-519	16L1189-16L1194	2-3/61	
H12-44	70	12L1111	3/61	Re# 301. Last FM switcher built
H16-44	520-525	16L1196-16L1201	1/63	
H16-44	603, 604	16L1202, 16L1203	2/63	CH-P 604 last FM locomotive built
H16-44	526-531	16L687-16L692	12/52	Ex-DL&W 930-935 (EL 1930-1935), acq. 1966
H16-44	532, 533	16L419, 16L425	7/51	Ex-NYC 7005, 7011, acq. from GE 1967
H24-66	534, 535	24L736, 24L743	6/53	Ex-DL&W 852, 859. Leased 9/69, acq. 1971

Chihuahua's chop shop

MEXICO'S CHIHUAHUA-PACIFIC definitely liked FM's H16-44, buying its first in 1955 and still asking for more when FM exited the locomotive business in 1963. Begun in 1900 as the Kansas City, Mexico & Orient, the CH-P did not complete its 671-mile main line from Ciudad Juarez (across the Rio Grande from El Paso, Texas) through the mountains to the seaport Topolobampo on Mexico's Pacific coast until December 1, 1961, and it needed the FM's for its newly opened line. All of the CH-P's H16-44s were built with high short hoods, and the five 600-series units were equipped for passenger service, as shown by the 602 (below) on the Los Mochis to Chihuahua *Vista Train* mini-domeliner on February 19, 1974. Sister 603 (opposite top) was riding the turntable at Ciudad Juarez in October 1971. In the mid-1960s, the CH-P embarked on a program to modernize its FM's and chopped the short hoods on about a dozen of its H16-44s at its La Junta shops— the only low-nose FM units ever produced. The 522 (opposite) showed the curved top of its new short hood at the depot in Ciudad Juarez in October 1971 before departing southward with mixed train No.52 for Madira. Boiling out black smoke (above) on January 2, 1979, the 521 looked more like an Alco than an FM. Blue smoke (below) is far more typical of an OP; nine CH-P H16-44s were rebuilt in-kind by United Railway Supply in Montreal in 1973-'74, including the 602.

PETER ARNOLD

Chihuahua gets the last

THWARTED IN ITS EFFORTS to buy more H16-44s by FM's decision to exit the locomotive business, the Chihuahua-Pacific turned to the used locomotive market in 1966 and snapped up the entire fleet of DL&W H16-44s (page 68) from the Erie Lackawanna. A year later it bought New York Central 7005 and 7011 (page 50), which had been traded-in to General Electric. These eight were the only Loewy carbody H16-44s on the CH-P. On February 19, 1974, the 531, ex-DL&W 1935, (bottom right) was smoking it up in Los Mochis, and the next day the 533, ex-NYC 7011, (below) was heading up a passenger train there. In 1969 the CH-P leased two former-DL&W Train Masters (page 70), purchasing them outright in 1971. Only the 534 (ex-DL&W 852) was repainted, and it is doubtful that either was used in revenue service; the 534 (opposite center) was in derelict condition at La Junta on March 10, 1974. They were the only Train Masters in Mexico.

Before indulging in its second-hand fleet, the CH-P scored a couple of significant "lasts" in Fairbanks-Morse production. In March 1961 it purchased the last switch engine produced by FM, H12-44 No.70 — and it was the only FM switcher in Mexico. Later renumbered 301 (left), it was working Chihuahua on September 5, 1973. And with the purchase of H16-44 604, builders number 16L1203, in February 1963, the CH-P acquired Beloit's last locomotive and went forever into the history books as FM's last railroad customer for a new locomotive. On February 19, 1974, the eleven-year-old 604 (opposite bottom) was working freight at Los Mochis. Today the CH-P fleet is almost entirely retired, although rumors persist that one or two might still be on the property. The high-nose 525 was put on display in Nuevo Casas Grande in 1991 in the new blue FNM livery of the unified national railways system. And in 1986, to mark the 25th anniversary of the completion of the CH-P through the mountains, the hulk of the chop-nosed 524 (bottom left) was installed high on the mountainside at Tremoris, shown on May 18, 1988, alongside a monument of inverted hoppers arranged in the outline of the state of Chihuahua.

BILL FARMER

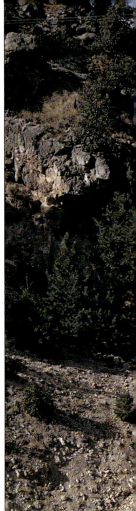

Doubleheading down to Cumbre

BOSQUES DE CHIHUAHUA			
MODEL	No.	B/N	DATE
H16-44	501	16L947	4/55
H16-44	1000	16L1195	12/61

IN THE MOUNTAINS of Chihuahua, 216 miles south of El Paso, Texas, the Bosques de Chihuahua (Chihuahua Forests) Railway operated a 20-mile line haul from the lumber mill at Mesa del Huracán to the Chihuahua Pacific interchange at Cumbre. The BdeCH operated its line with two H16-44s painted in a livery inspired by the Train Master demonstrators (page 6). Since the units had no multiple-unit capability, they would doublehead with two engineers on the road job. On March 12, 1974, the 1000 was leading the 501 (above) down the steep grade into Cumbre before working the CH-P interchange (opposite bottom). The units were in reverse order (opposite top) earlier that same day at Mesa del Huracán. On March 20, 1975, the 501 (below) was dropping downgrade into Cumbre (which has no highway access) as the CH-P mixed train appeared behind a pair of H16-44s led by 509. A passenger excursion was run on the BdeCH on May 14, 1988, and the line was abandoned shortly thereafter. The two H16-44s were "moved to another operation," but in recent years no one can seem to confirm their fate.

Second-hand survivors

MIKE SCHAFER

1984, and his OP empire collapsed.

As the Chicago & North Western was retiring its Baby Train Masters, Precision National Corporation acquired the 1677 and 1699 on a one-year lease in October 1975. The 1677 was then leased briefly to Peabody Coal and returned to PNC, where it was photographed (below) in Chicago on September 23, 1976; both units were scrapped shortly thereafter. Peabody also leased from the Frisco H10-44 274 (opposite center right) for use at its River King No.1 Mine in Freeburg, Ill., where it was working in December 1974.

Border Steel of Anthony, Texas, and its affiliated Metal Processing Inc. were responsible for cutting up many Western FM's, but they also employed ex-Santa Fe H12-44s 557 as No.99

A SURPRISINGLY FEW Fairbanks-Morse units made it onto the used locomotive market, primarily because their OP engines were so valuable in non-railroad applications that they were often sold off as the locomotives were retired and scrapped. The exceptions, though, were fascinating.

One of the last major railroads to keep its FM fleet was the Milwaukee, and in the early 1980s it sold nearly a dozen switchers to John Zerbel for use on a 400-mile collection of branch lines he was operating as the Central Wisconsin and Wisconsin Western. Only a few received the company's blue livery, shown (above) on WIWR 1204 (ex-Milwaukee H10-44 767), alongside sister 763 at Monroe, Wis., in September 1982. Mr. Zerbel died in

PAUL HUNNELL

KEITH E. ARDINGER

and 560 as MPI 100. In April 1978 they were both working the steel mill (opposite top) with the 99 numbered as Bicentennial 1776 (inset). The 100 is today at the museum in Perris, Cal.

Monon's H10-44 No.18 was re-engined EMD in 1961 (page 23) and wound up working for W.R. Grace Chemical in Mulberry, Florida, where it was photographed (left) on April 21, 1968, as second 102, while the third 102, a brand new SW1500, worked in the background. The FM had inherited the trucks from first 102, an ex-ACL Baldwin VO1000.

Still at work in mid-1996 is Hallett Dock Co. HD-11, photographed (right) at West Duluth, Minn., on May 9, 1980. This unit was Minneapolis, Northfield & Southern 11 (page 24). Its mate, MN&S 10, is displayed as "C&NW 10" at Milton Junction, Wisconsin.

TWO PHOTOS: "J.W.P." / BOB YANOSEY COLLECTION   KERMIT GEARY, SR.

GARY L. POWELL / MATT HERSON COLLECTION

Honorably discharged

IN JANUARY 1953 the United States Army Transportation Corps took delivery of 20 H12-44s and distributed them to bases all across the continental U.S. They all had spartan carbodies and uniform Army black paint with yellow trim. As they were maintained over the years — generally to pristine military standards — the units picked up details of individuality like variations in stripes and additional outboard handrails. The 1858 (opposite right) was at Ft. Carson, Colo., on January 30, 1974, wearing a Transportation Corps herald on its hood and a set of custom handrails. The 1846 (opposite top) and sister 1849, working at Ft. Knox, Ky., in May 1979 alongside the Illinois Central Gulf Paducah-Louisville line, had "fake H10-44" sheet metal cab roof overhangs applied. Adjacent to the 1846's builders plate (opposite bottom left) was an Army specification plate. By the early 1980s the Army had the largest intact fleet of FM units in the country, and thanks to federal regulations on the disposition of military property, most were eligible for donation to non-profit organizations. As a result, fully half of the fleet has been preserved in museums, and most are still in operating condition! The chart at right is a true honor roll of veterans.

The 1850 (below) was in service at the Heart of Dixie Railroad Museum in Birmingham, Ala., in February 1996 along with sister 1861; both had come from the Anniston Army Depot. The 1858 had migrated from Ft. Collins to the Sunny Point Military Ocean Terminal in North Carolina before going to the North Carolina Ports Railway, where it will be salvaged for parts to keep sisters 1852 and 1860 (page 79) in service. The 1846 and 1849 from Ft. Knox wound up at the Kentucky Railway Museum and Bluegrass Railroad Museum, respectively.

The table (opposite) lists all of the FM units known to have survived into 1996 — and the Army switchers make up more than a third of the group. Only ex-Army LTS 1845, Hallett Dock HD-11, the North Carolina Ports fleet and the TVA Gallatin Baby Train Master are in revenue service. But the very first FM locomotive, Milwaukee Road 760, still runs (page 128) at the Illinois Railway Museum.

DISPOSITION OF U.S. ARMY H12-44s, BUILT 1/53

Number	B/N	Location and Notes
1843	12L667	**Rochester, N.Y.**; Rochester NRHS museum; from Seneca Ordinance Depot, N.Y.
1844	12L668	**Kirkwood, Mo.**; serviceable at National Railway Museum; from Seneca OD, N.Y.
1845	12L669	**Morrisville, Pa.**; LTS 1845; in service USX Fairless; from Radford Ordinance Depot, Va.
1846	12L670	**New Haven, Ky.**; in service on Kentucky Ry. Museum; from Ft. Knox, Ky.
1847	12L671	**San Francisco, Cal.**, Golden Gate RR Museum; from Hawthorne Munitions Plant, Nev.
1848	12L672	**McAlester, Okla.**; McAlester Army Depot 8/96 for sale
1849	12L673	**Versalles, Ky.**; Bluegrass RR Museum; from Ft. Knox, Ky.
1850	12L674	**Birmingham, Ala.**; Heart of Dixie RR Museum; from Anniston Army Depot, Ala.
1851	12L675	**Ogden, Utah**; last reported stored at Hill AFB; present status unknown
1852	12L676	**Morehead City, N.C.**; being overhauled N.C. Ports Ry; from Sunny Point MOT, N.C.
1853	12L677	**Anniston, Ala.**; last reported stored at Anniston Army Depot; status unknown
1854	12L678	**Ogden, Utah**; Defense Supply Agency 53205; status unknown
1855	12L679	**Boulder City, Nev.**; Nevada State RR Museum; from Sierra Weapons Depot, Cal.
1856	12L680	**Fremont, Cal.**; active on Niles Canyon Scenic Ry; from Oakland Army Terminal
1857	12L681	**Portola, Cal.**; active at Feather River Rail Society museum; from Sierra WD, Cal.
1858	12L682	**Morehead City, N.C.**; parts unit for N.C. Ports Ry; from Sunny Point MOT, N.C.
1859	12L683	**Ft. Wingate, N.M.**; listed as surplus at Ft. Wingate Army Depot; status unknown
1860	12L684	**Morehead City, N.C.**; in service on N.C. Ports Ry; from Sunny Point MOT, N.C.
1861	12L685	**Birmingham, Ala.**; in service Heart of Dixie Museum; from Anniston Army Depot, Ala.
1862	12L686	**McAlester, Okla.**; sold Army Depot 4/9/96 to scrapper L. Mihalakis, scrapped on site

Rosters compiled by Wayne Monger and Jim Boyd

JOHN P. LYLE III

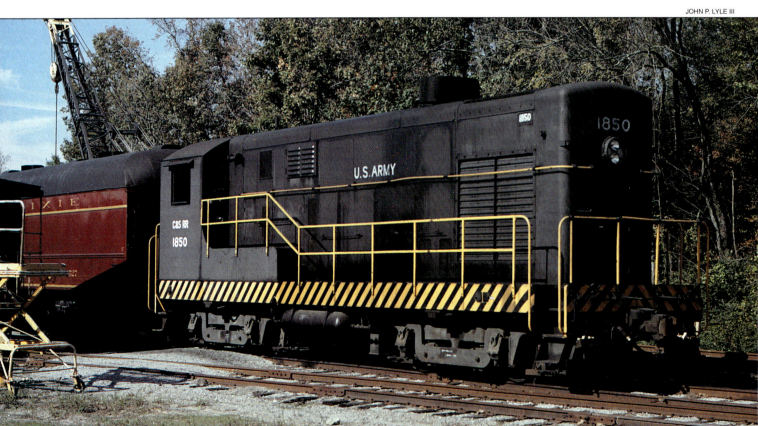

BOB YANOSEY COLLECTION

SURVIVING FM LOCOMOTIVES: 1996

MODEL	RR AND NUMBER	B/N	DATE	LOCATION AND NOTES
H10-44	Milwaukee Road 760	L1001	8/44	**Union, Ill.**; Illinois Railway Museum, in service
H10-44	Hallett Dock 11	L1019	9/46	**Duluth, Minn.**; Ex-MN&S 11, nee-MW 51; in service
H20-44	Union Pacific DS-1366	L1032	8/47	**Union, Ill.**; IRM; ex-Southwest Portland Cement 409, nee-FM demo 2000
H20-44	Union Pacific DS-1369	L1042	12/47	**Campo, Cal.**; San Diego RR Museum; ex-SWPC 408
H10-44	Apache 200	10L100	11/48	**Centerville, Tenn.**; wrecked on South Central Tennessee, hulk reported still at wreck site
H10-44	Milwaukee Road 781	10L328	3/50	**Brodhead, Wis.**; on display
H12-44	MN&S 10	12L427	1/51	**Milton, Wis.**; Nee-MW 10; carbody displayed at Liberty Station Restaurant as "C&NW 10"
H12-44	Weyerhaeuser Timber 1	12L437	8/51	**Snoqualmie Falls, Wash.**; Pacific Transp. Services 121; preserved PS&SV Ry museum
H12-44	AT&SF 608	12L439	10/51	**Sacramento, Cal.**; preserved at California Ry Museum; nee-AT&SF 508
CPA16-4	CPR 4065	CLC 2647	12/51	**Ottawa, Ontario**; displayed at National Museum of Science & Technology
H12-44	USS (Fairless) 26	12L613	2/52	**Morrisville, Pa.**; stored derelict in steel mill yard
H12-44	Milwaukee Road 731	12L562	3/52	**Belton, Mo.**; carbody preserved at Kansas City Ry Museum
H12-44	TVA 22	12L778	3/52	**Morehead City, N.C.**; stored 6/96 Pickens, S.C.; parts unit for N.C. Ports Railway
H12-44	US Army 1843-1862	12L667-86	1/53	**See adjacent chart, page 124**
CFB16-4	CPR 4455	CLC 2727	4/53	**Cranbrook, B.C.**; Cranbrook Museum of Rail & Travel; ex-BC Rail radio robot car RCC3
CFB16-4	CPR 4456	CLC 2728	4/53	**Cranbrook, B.C.**; Cranbrook Museum of Rail & Travel; ex-BC Rail radio robot car RCC4
H12-44	Yankeetown Docks 1	12L776	8/53	**Morehead City, N.C.**; at Pickens, S.C., being overhauled as N.C. Ports Ry 1801
H12-44	Milwaukee Road 740	12L823	1/54	**Bellevue, Ohio**; carbody displayed at Mad River & NKP Museum
H20-44	AC&Y 505	20L832	3/54	**Galveston, Texas**; Center for Transp. & Commerce; ex-SWPC 410; painted as "UP 410"
CPA16-4	CPR 4104	CLC 2726	4/54	**High River, Alberta**; preserved at Museum & Highwood Railway Project
H16-44	CPR 8554	CLC 2897	7/55	**Calgary, Alberta**; stored for preservation
H12-44TS	AT&SF 543	12L1023	5/56	**Sacramento, Cal.**; displayed at California State Ry Museum
H12-44	Yankeetown Docks 2	12L1024	5/56	**Morehead City, N.C.**; at Pickens, S.C., being overhauled as N.C. Ports Ry 1802
H24-66	CPR 8905	CLC 2922	6/56	**St. Constant, Quebec**; displayed at Canadian Ry Museum; only surviving Train Master
H12-44	SP 1581	12L1066	10/56	**Dallas, Texas**; ex-SWPC 44; preserved at Age of Steam Museum
H12-44	AT&SF 560	12L1095	4/57	**Perris, Cal.**; ex-MPI 100, preserved at Orange Empire Ry Museum
H12-44	Ayrshire Collieries 1	12L1100	4/57	**Morehead City, N.C.**; Yankeetown Docks 3; at Pickens, S.C.; parts unit for N.C. Ports Ry
H16-66	TVA 24	16L1157	10/58	**Gallatin, Tenn.**; in service!
H16-44	CH-P 524	16L1200	1/63	**Tremoris, Ch., Mexico**; displayed on site of last spike on CH-P
H16-44	CH-P 525	16L1201	1/63	**Nueva Casas Grande, Mexico**; on display in NdeM blue

The FM's of the Fairless Works

BACK IN 1951 United States Steel needed a new state-of-the-art steel mill to exploit the East Coast and Atlantic export markets, and the result was the 3900-acre "Fairless Works" near Morrisville, Pa., northeast of Philadelphia. The extensive rail network feeding the mill was initially equipped with sixteen 1200-h.p. diesel switchers, eight Baldwin S12s and eight Fairbanks-Morse H12-44s (GE9-GE16, the GE standing for "General Equipment"). Over the years the Fairless fleet expanded with the acquisition of new and used locomotives, including five more H12-44s, two ex-PRR, two ex-NYC and ex-U.S. Army 1845. Since the Fairless Works is on a peninsula in a bend of the Delaware River and surrounded on

TWO PHOTOS: KERMIT GEARY, JR.

UNITED STATES STEEL - FAIRLESS WORKS

MODEL	NUMBER	BUILT AS	B/N	DATE	TO USS	NOTES
H12-44	GE9-GE16	Same	12L571-12L578	11/51-2/52	—	All scrapped
H12-44	23	PRR 8721	12L647	11/52	1970	Ex-PC/PRR 8337, scrapped
H12-44	24	PRR 8714	12L640	11/52	1970	Ex-PC/PRR 8330, scrapped
H12-44	25	NYC 9120	12L837	1/51	1970	Ex-PC/NYC 8309, scrapped
H12-44	26	NYC 9121	12L613	2/52	1970	Ex-PC/NYC 8310, hulk on property 5/96
H12-44	27	US Army 1845	12L669	1/53	1972	Sold to Locomotive Trouble Shooters 1991, Re# 1845, in service Tyburn R.R 5/96

three sides by water and is fenced for tight security, the FM's worked out their years in near total obscurity. The GE9 and former-PRR 23 (opposite bottom left) were photographed in retirement in September 1982 and January 1984, respectively. U.S. Steel 26, the ex-NYC 9121 (opposite top) was sitting derailed in an unused yard beside the mill on April 11, 1996.

One FM "escaped" from Fairless, only to return there. Army 1845, USS 27, was sold to Mike Crane of Locomotive Troubleshooting, who gave it a coat of blue paint and its Army number and put it to work just outside the gate of the Fairless Works, first at an A.H. Staley feed mill and then at the Tyburn Railroad's bulk-transfer facility (above), where on April 11, 1996, it was switching covered hoppers of soda ash which was then trucked to the nearby Dial Soap factory. The H12-44, with its fireman's window still blanked over from its Fairless "hot side" duties, was spotting a hopper as the Conrail local (top) pulled the day's traffic out of the Fairless Works. A few months later the FM went back behind the gates as LTS got a switching contract to work the mill. ⊕

FLASH ASSISTANCE BY LLOYD RINEHART and GREG PLATT

Firsts and lasts

THE VERY FIRST FM locomotive ever built, Milwaukee Road H10-44 760, survives in operating condition at the Illinois Railway Museum in Union, Illinois. On August 25, 1996, the 760 pulled Southwest Portland Cement H20-44 409 out for a night photo session, specifically for this book. The presently unserviceable 409 was built in August 1947 as FM demonstrator 2000, before becoming Union Pacific 1366. On August 24, 1996, the Milwaukee Road's first spartan carbody H12-44, 740 (built in February 1954 as 2310), was on display (right) at the Mad River & NKP Museum at Bellevue, Ohio. The sole surviving Train Master, Canadian Pacific 8905 (below), was posed at the Canadian Railway Museum in Delson, Quebec, alongside CPR's first diesel, the National Steel Car 7000, powered by a Harland & Wolff engine from Northern Ireland. Museums have saved many of FM's firsts and lasts.

MIKE DEL VECCHIO

128